Railway World **SPECIAL**

THE WELSH NARROW GAUGE RAILWAYS

PETER JOHNSON

LONDON

IAN ALLAN LTD

First published 1985
Second Edition 1991

ISBN 0 7110 1977 0

Published by Ian Allan Ltd, Shepperton, Surrey; and printed by Ian Allan Printing Ltd at their works at Coombelands in Runnymede, England

Dedication

This book is dedicated to my Mother, who died 11 September 1988, with fond remembrance. She enjoyed visiting several of the lines described herein and always supported my interest in them.

Placenames

The anglicised form of Welsh placenames common in the 1950s has been used when referring to that period or earlier; otherwise the current Welsh forms have been used.

The Festiniog Railway Company was spelled so in the 1832 Act of Parliament. To bring it into line with the placenames would require another Act, and considerable expense. The company has taken to using the form 'Ffestiniog Railway' as a trading name in recent years.

Contents

Right:
Llangower, the intermediate crossing place on the Bala Lake Railway on 19 August 1989. The locomotive is *Maid Marian.*

Previous page:
On its way up Snowdon No 9 *Ninian* waits at Hebron for a Down train to cross on 18 August 1989.

Front cover:
Ffestiniog Fairlie, *Earl of Merioneth*, skirts round Llyn Ystradau soon after commencing its journey to Porthmadog. *Peter Johnson*

Back cover, top:
British Rail's Cambrian Coast line is just visible in this photograph of *Russell* on Gelerts Farm Curve on the Welsh Highland Railway in August 1989.
D. W. Allan

Back cover, bottom:
***Maid Marian* at Llangower on the Bala Lake Railway on 13 August 1989. The record books state that the locomotive was built in 1903 but when it was realised that it had the frames of the original *Alice*, built in 1889, the railway staged a centenary celebration during its Summa Gala on 24/25 June 1989.**
Tom Heavyside

Preface

Since the first edition of this book appeared in 1985 the Welsh narrow gauge lines have continued to develop, often in ways which could not have been foreseen – the sale of the vale of Rheidol Railway to private enterprise by British Rail had often been the subject of discussion and speculation but who could have guessed that it would be taken over by an existing independent railway company?

However, our review of these lines begins with the situation as it was in the 1950s, when the tourist had only the Great Orme Railway, the Talyllyn Railway and the Vale of Rheidol Railway and, by the end of the decade, the Festiniog Railway to ride on.

The Talyllyn and Festiniog were functioning under new management (from 1951 and 1955 respectively) with volunteer assistance and were to establish the trends for much of what was to happen afterwards, both in Wales and in railway restoration and preservation elsewhere. At the same time, until 1956, the 2ft 6in gauge Welshpool & Llanfair Light Railway was still being operated as a goods line by British Railways, the 4ft gauge Padern Railway was still being used to carry slate from the Dinorwic quarries to Port Dinorwic using the trackbed later to be occupied, in part, by the Llanberis Lake Railway (1ft 11½in gauge from 1971); and the formations now used by the Bala Lake (2ft gauge from 1972), Brecon Mountain (60cm gauge from 1980), Teifi Valley (1ft 11½in gauge from 1986) and Welsh Highland railways (2ft gauge from 1980) were still being used by British Railways' standard gauge traffic.

The 1960s were a boom time for the existing lines; increasing patronage and the fresh availability of otherwise unused trackbeds were to provide the impetus for the construction of new narrow gauge railways intended to cater solely for the tourist trade from the outset. This development was to extend the geographical range of the lines into South Wales, too.

Of the 15in gauge lines only the Fairbourne Railway has operated continuously, although it is now a 12¼in gauge railway. The Rhyl Marine Lake Miniature Railway operated until 1969, when it was closed and dismantled; it was reestablished on the same site under new management 10 years later. The Rhiw Valley Railway was established privately in the 1970s and enjoyed a brief period of public operation in the 1980s following a request from the Wales tourist Board.

The rope-hauled Great Orme Railway also has a record of continuous operation despite having reverted to being a tramway. As a rope-hauled system it was joined by the 1 in 1.8 incline railway at Llechwedd Slate Caverns, Blaenau Ffestiniog in 1980. Llechwedd is also the home of the miners' tramway which is another means of exploring the underground caverns, excavated in the quest for slate, which exist on the site.

The establishment of museums with a greater or lesser interest in the Welsh narrow gauge railways is a development of the postwar era. Both the Narow Gauge Railway Museum, at the Talyllyn Railway's Wharf station, the Festiniog Railway Museum, at Portmadoc Harbour station, were established in 1956. They were followed by the Industrial Railway Museum at Penrhyn Castle, Bangor, which opened in 1965. During the 1970s museums were opened at Gilfach Ddu, Llanberis; Gloddfa Ganol, the former Oakeley Quarry site at Blaenau Ffestiniog; Corris and Betws-y-Coed. Following the current trend for museums to be interactive these four all have some sort of working line, either operational or planned, on a regular or casual basis.

In 1970 the major lines established the Narow Gauge Railways of Wales Joint Marketing Panel to promote themselves. The activities of the panel manifest themselves in the slogan, 'Great Little Trains of Wales', joint timetables and posters, and the Narrow Gauge Tourist Ticket which allows unlimited travel on member lines for seven days. The members undertake to distribute each others' publicity in their respective localities. The result is a high level of publicity which would not be obtainable by any of the railways working independently.

The photographs in this book are intended to show something of these 'Great Little Trains' during recent years, with the occasional backwards glance. It has not been possible to illustrate every happening on every railway but no apology is made if the reader does detect a bias towards the Festiniog Railway; it is the oldest of the lines covered, having celebrated its 150th anniversary in 1986, and has a distinguished record of continuous and ongoing achievement. It is also the railway the author knows best!

Acknowledgements

My sincere thanks are extended to Peter Miller and Ray Sanderson for their contribution to the text; any errors escaping their attention must, of course, be laid firmly at my door.

I hereby express my thanks to all the railways for providing, whether wittingly or not, facilities for photography. Most had personnel who readily spoke about their lines and who willingly increased my knowledge and enjoyment of them; to those individuals, many of them unknown to me by name, I express my thanks too.

I am also most grateful to the photographers, especially those who submitted their work to me for consideration; I am sorry that I could not make greater use of their efforts. The fine summer of 1989 undoubtedly made my task easier, as will be seen from the many selected views taken that year.

If you bought this book whilst visiting the 'Great Little Trains', it is my wish that it aids your enjoyment of them; otherwise I hope you are encouraged to visit, or revisit, them.

Peter Johnson
Leicester

Photographs are by the author except where stated.

THE WELSH NARROW GAUGE RAILWAYS

Introduction

The charm of narrow gauge railways has long been recognised, by both enthusiasts and passengers alike, but to identify the cause of the attraction is not so easy. Possibly it arises from the combination of the many ordinary facets which go to make up any railway and which become distinctive when applied to the narrow gauge. For example, the locomotives may be diminutive or unique; the carriages could be of great age (and memorably uncomfortable!) or home-made; the station buildings, where they exist, might be little more than garden sheds, the platforms overgrown or non-existent; the journeys often slow, the countryside remote; the scenery, in almost all cases, superb. All these factors can be found elsewhere, but many of them are exclusive to narrow gauge railways and on the last count, that of superb scenery, it is not surprising that the greatest concentration of narrow gauge railways in these islands.and occurs in the spectacular scenic grandeur that is Wales or, more specifically, in that part of North Wales known as Snowdonia. There are exceptions of course, but most of the lines mentioned in this volume are to be found in a compact area which makes them a popular tourist attraction, as much part of a Welsh holiday or excursion as the mountains, lakes and castles. They are as Welsh, too, as bara brith and lava bread despite the disparity of their origins, which often has little to do with the surroundings in which they operate; in fact it is the contradictions which make them essentially Welsh.

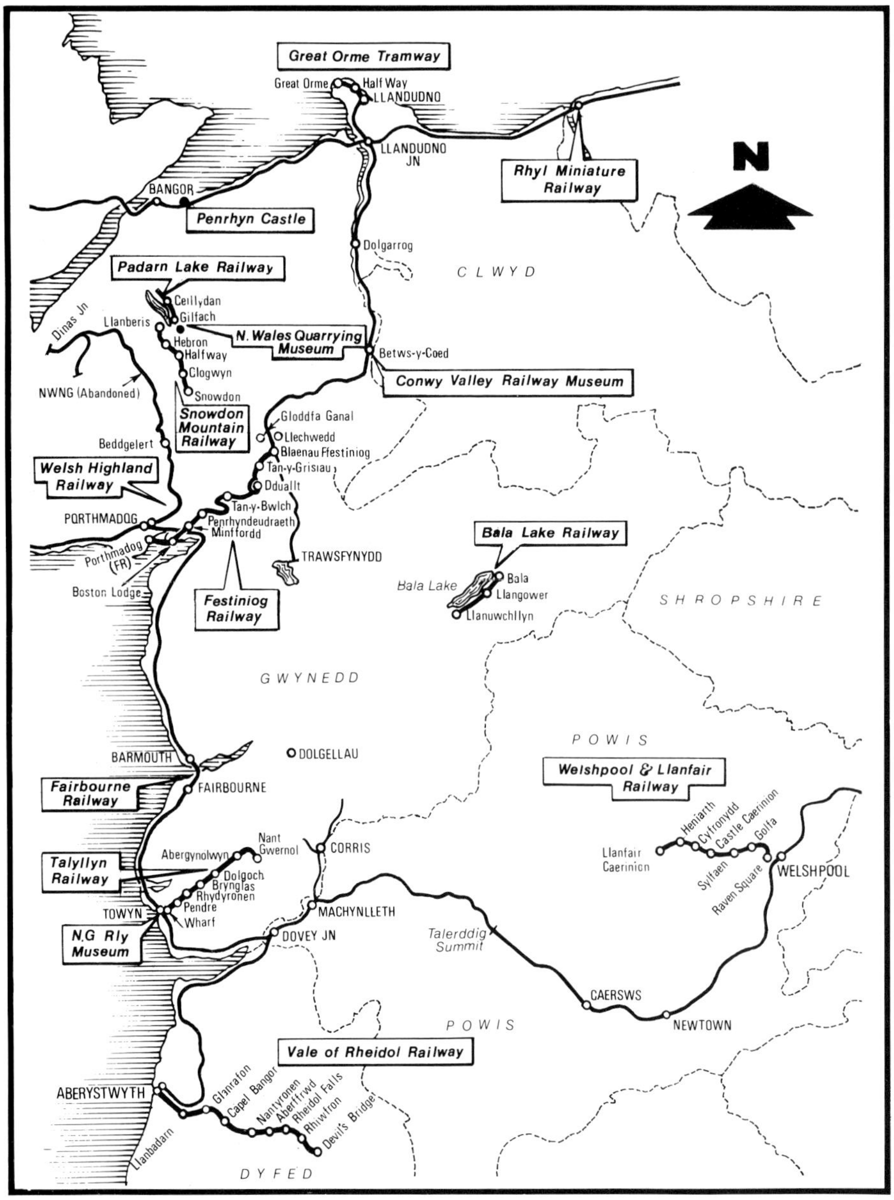

Bottom:
In its third year of operating the revived Festiniog Railway's track still showed much evidence of the closure period. *Prince* approaches Minffordd bound for Penrhyn in July 1957. *Peter Treloar*

Above right:
It was often claimed that before preservation the Talyllyn track was only held together by the turf which surrounded it, certainly there is little evidence of sleepers in this photograph of *Dolgoch*, then the railway's only working locomotive, near Abergynolwyn on 12 September 1947. *S. W. Baker*

Top right:
Great Western Railway No 822 *(The Earl)* with a mixed goods train close to Raven Square on the now-abandoned section of the Welshpool & Llanfair Light Railway on 25 March 1947.
R. E. Tustin

Above right:
On the Snowdon Mountain Railway the early morning shed scene is almost timeless. This August 1937 view could easily be 10 years earlier or 20 years later. *R. E. Tustin*

Right:
The fine water tower was one item of equipment which did not survive the Vale of Rheidol Railway transfer to the former standard gauge shed in 1968. On 12 August 1957 No 8 *Llewelyn* was being prepared to work the 2.30pm departure from the 1925 station. The former Harbour branch ran to the left of the water tower and sheds. *Geoffrey Bannister*

Bottom right:
An untypical view of the Fairbourne Railway on a summer's day in 1954, with *Gwril*, the Lister tractor, providing the traction for a train leaving Penrhyn Point for Fairbourne. *Michael E. Ware*

Festiniog Railway

The Festiniog Railway Company gained its act of incorporation in 1832 and opened its line between Portmadoc and Blaenau Ffestiniog in 1836. It was built to carry slate from the quarries at Blaenau Ffestiniog to the harbour at Portmadoc – including the branches to Dinas and Duffws at Blaenau, a distance of just over 14 miles. A narrow gauge of 1ft 11½in was adopted and the railway was designed to have a continuous falling gradient down to the harbour so that laden trains could be worked down by gravity; horses pulled the empty wagons back up the hill. The last mile to Portmadoc utilised Madock's 1811 embankment across the Glaslyn estuary – the Cob. Adapting the continuous gradient to the contours put the railway onto some very exposed ledges and also resulted in numerous curves. The sharpest of these, called Tyler's after the Board of Trade inspector, is 2½ chains radius. Where the contours were unsuitable some imposing dry-stone embankments were built; the highest, at Cei Mawr, is 62ft high. To improve the route the Moelwyn Tunnel (730yd) was opened in 1842, followed by the Garnedd Tunnel (60yd) in 1851; the length and narrow bore of the former, combined with a lack of ventilation, were to result in the railway achieving a certain notoriety, especially when the 20th century's inter-war years bought an increasing number of tourists as passengers.

The demand for slate in the developing industrial areas of England and elsewhere brought much traffic to the railway which was soon operating to capacity. To overcome the difficulties encountered the track was modernised and in 1863 steam locomotives were introduced. Initially there were four 0-4-0 tank engines built in London; they were followed in 1867 by two similar, but larger, machines from the same source. Greatly rebuilt and modified, four of them still survive, although only one is presently in working order.

The introduction of steam engines led the way to the (official) carriage of passengers, a move which was encouraged by the threat of competing routes to Blaenau Ffestiniog. Using tiny four-wheeled carriages built in Birmingham the first passenger trains ran in 1865.

Increasing traffic again brought operating problems which the company sought to solve by laying a second track. Parliamentary powers had been obtained and the necessary land purchases put in hand when the problem was solved by the introduction of a locomotive capable of prodigious feats of haulage when compared with the existing fleet.

The saviour was *Little Wonder*, an articulated locomotive built, also in London, to Robert Fairlie's patent. Delivered in 1869, it was fitted with a double-ended boiler unit having twin fireboxes in the middle. The boiler was carried on two four-coupled power bogies which could swivel under it, the crew being carried one on either side of the fireboxes. The resulting machine was effectively twice the size of the original engines but, being articulated, was still able to cope with the railway's bends and gradients. *Little Wonder* proved to be very successful, attracting attention from all over the world. The inventor gave the company free use of the patent and the double Fairlie became the Festiniog's trademark.

Having proved that articulation worked on a locomotive, the company then applied the same principle to its passenger stock in a move to increase both capacity and comfort. The year 1872 saw the introduction of two iron-framed bogie carriages, most likely the first of this type to run in the British Isles. These vehicles, Nos 15 and 16, are still in regular service.

During the last years of the 19th century the Festiniog Railway reached its peak, culminating in the construction, in 1879 and 1885, of two double engines, *Merddin Emrys* and *Livingston Thompson*, in its own Boston Lodge works. These locomotives were designed by George Percival Spooner, whose father, Charles Easton Spooner, had an inventive streak which had helped bring the railway to public attention. In turn, it had been Charles Easton's father, James Spooner, who participated in surveying the railway's route before construction began. It is no coincidence that the era of the Festiniog Railway's pre-eminence in the field of narrow gauge railways coincided with the period in which the Spooners were involved with the railway.

With the arrival of the 20th century the FR was to enter a period of decline, despite greater use of the line by tourists, which was to end, in 1946, in closure and abandonment. The decline was caused both by the tapping of the Blaenau slate traffic by the LNWR (from 1881) and the GWR (from 1883) and the development of alternative roofing materials.

The only bright spot arose during the 1920s; the number of passengers carried was considerably increased by tourists seeking to view the remote scenic splendours visible from the 'Toy Railway' or 'Faery Line', as the company promoted itself, whilst they passed through the Vale of Ffestiniog. Even this traffic was to

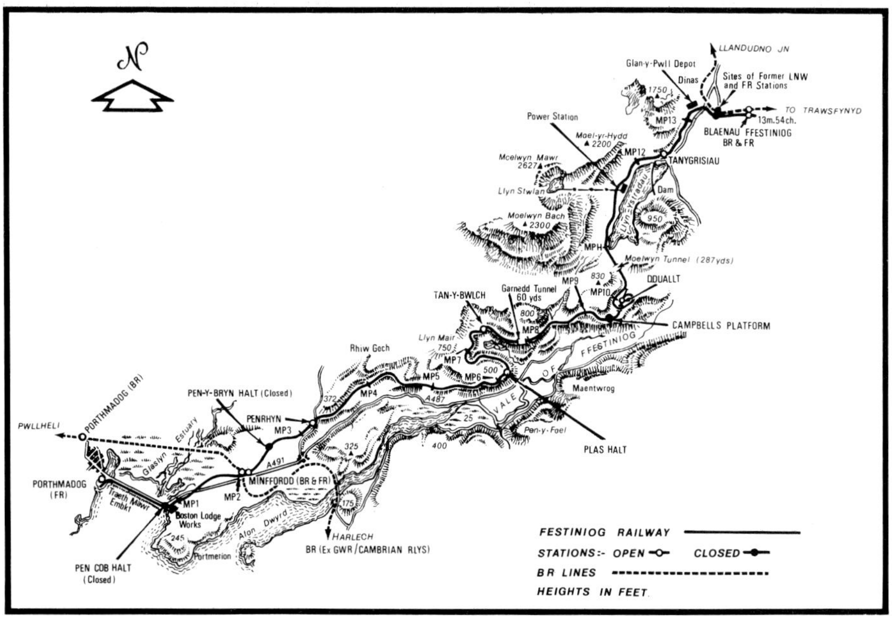

Above:
The 'new' Festiniog Railway in Blaenau Ffestiniog. *Mountaineer*, built by the American Locomotive Co in 1917 for service on the War Department Light Railways in France and donated to the FR in 1967; it is seen at the modernised Glan-y-pwll level crossing on 8 September 1984, two years after the Tanygrisiau-Blaenau Ffestiniog section was restored and reopened. *Tom Heavyside*

decline during the 1930s, despite efforts to attract new business. Engine failures could easily make travel on the 'Faery Line' a journey of epic proportions, and missed connections with the standard gauge links were commonplace. Meanwhile, the quarrymen took to the buses.

In 1923 the narrow gauge Welsh Highland Railway was opened from Portmadoc to Dinas, near Carnarvon, via Aberglaslyn, a distance of 22 miles. It was never successful and closed on 31 December 1933. The FR took a lease on its neighbour and operated a summer service during the following three years, though with no greater success. At the end of the 1936 season the line was closed for good.

From 1924, until he died in 1931, the railway came under the influence of the light railway king, Col Holman F. Stephens. The Colonel was well known for the ramshackle collection of minor railways he ran, with varying degrees of success, from an office in Tonbridge, Kent. In addition to being a member of the FR Company board, he was also appointed Engineer, Managing Director, Locomotive Superintendent and Company Chairman at various times, sometimes concurrently. Stephens was also involved with the Welsh Highland Railway at the same time. Unfortunately his rather brusque approach was not well received by spirited Welsh railwaymen and the two sides rarely agreed, if at all, on the best way of achieving the desired result.

On the FR the passenger service was withdrawn on the outbreak of war in 1939. Slate trains continued to run as required, and when motive power was available, throughout the war. Afterwards it quickly became clear that there were no resources, either within the railway or elsewhere, to repair the track and rolling stock so in 1946 the Festiniog Railway was closed.

Powers to close the line legally could only be had by spending money the company did not have to promote an Act of Parliament. Everything was left to rot, or to be vandalised, where it stood.

Almost immediately attempts were made to find ways of reopening the railway, possibly using volunteers to assist a nucleus of paid staff. Eight years were to pass before all the legal and financial obstacles were overcome and a group of people, including Alan Pegler, provider of the necessary financial support, were able to implement a new regime. They were

Below:
The 1863-built *Prince*, now in 1900s livery, arrives at Minffordd station with the Vintage Train on 20 April 1986, during the railway's 150th anniversary celebrations.

Left:
Ex-Penrhyn Slate Quarry Railway *Blanche* pulling up at Minffordd with the 'Mountain Prince' on 21 August 1987. This station has been the centre of much volunteer activity by the Railway's Parks & Gardens Department.

supported by the newly formed Festiniog Railway Society.

During the closure much of the track between Boston Lodge and Tan-y-Bwlch had become overgrown so there was no prospect of running trains throughout immediately. Sand, blown in from the beach was cleared from the track at Boston Lodge so that a service could be operated across the Cob. The following year, on 19 July 1955 the new management opened for business, carrying 22,000 passengers by the end of the season. The ex-WD Simplex, now named *Mary Ann*, was used at first but *Prince* became available at the beginning of August and was used exclusively until 1957.

In 1956 the service was extended to Minffordd, in 1957 to Penrhyn and in 1958 to Tan-y-Bwlch, 7½ miles from Portmadoc. From the 1957 season *Prince* was joined by *Taliesin*, the 1885-built Fairlie which was originally named *Livingston Thompson*.

As the Festiniog revival was getting under way work was starting on the pumped storage electricity scheme at Tanygrisiau. This involved flooding the FR to the north of the Moelwyn Tunnel. The Company made representations in protest, going as far as the House of Lords, but to no avail; railway preservation and restoration was an unknown quantity in the 1950s and it took until 1971 before compensation was agreed.

Meanwhile, the railway consolidated its achievements and by 1968, when it was extended a further two miles to Dduallt, it had: returned the other double Fairlie, *Merddin Emrys*, to traffic; acquired Hunslet *Linda* and *Blanche* and several miles of rail from Penrhyn Quarry; been given the American-built *Mountaineer*; bought Planet diesel *Upnor Castle* from the Welshpool & Llanfair Railway; put nine bogie and five four-wheel FR coaches into service; acquired and rebuilt a Lynton & Barnstaple Railway coach, now buffet car No 14, from Devon; rebuilt ex-Welsh Highland Railway carriage No 26 which had been used as a chicken house at Groeslon; and built five new bogie carriages.

Dduallt is one of those rare stations with no road access, passengers alighting there could see, for the first time, the results of the efforts to build a new route to Tanygrisiau which had commenced in 1965. After a great deal of discussion and negotiation, and following several surveys, a route had been chosen which showed that the Festiniog Railway of the 20th century could be as adventurous as that of the 19th century. The new line, or deviation, had to gain height, to bypass the CEGB's lower reservoir without introducing any excessive gradients. This condition was met by the construction of a spiral around the hill at Dduallt. This work, and much else besides, was carried out by groups of volunteers who came to call themselves 'Deviationists'! The whole project was to take 13 years to bring to fruition, often with minimal resources.

In addition to the spiral, the new route still required some major earthworks, including a new (294yd) Moelwyn Tunnel. At the CEGB's insistence those at the rear of the Tanygrisiau power station, including underground bridges over high-pressure pipelines, had to be carried out by contractor, but the railway was able to use its own labour on the tunnel. Three miners, with experience of Cornish tin and South African diamond mines, were hired for the job. Construction started in September 1975 and took eight months, although finishing work and track laying meant that it was 1977 before passenger trains passed through.

In that year a temporary station, called Llyn Ystradau and 11¼ miles from Porthmadog, was opened alongside the lake to satisfy the conditions attached to the grants received towards the cost of making the tunnel. At the same time outside funding became available to provide labour, in the form of a Manpower Services Commission scheme, for the contractor to build the power station section. The 'Deviationists', meanwhile, had transferred their efforts to the site of the new station at Tanygrisiau. Twelve miles from Porthmadog, it was opened on 24 June 1978.

A further four years were to elapse before the FR completed its restoration to Blaenau Ffestiniog. To finish the final 1½ miles it was necessary to build bridges over the Cwmorthin road at Tanygrisiau and the Afon Barlwyd at Glan-y-pwll; to rebuild and raise four footbridges over the line (modern rolling stock is higher than the original); underpin embankments; restore walls and fences; and to establish a site for a terminus.

This work was undertaken by a mixture of volunteers, company staff and job creation labour. Some of the volunteers undertook direct responsibility for particular aspects of the work and were to become specialists in their fields – bridges, culverts, dry-stone walling, for example. Material and equipment was stored at a depot established on the site of the old engine shed by the level crossing at Glan-y-pwll.

After considering the site of the former LNWR exchange station, opposite the BR Blaenau Ffestiniog station, for a terminus, the FR was persuaded to participate in a local authority scheme to build a new joint station in the town. Funding to clear the site, make the necessary road alterations for the bridges which gave the FR access to the site, and for car parks came from several government and non-government bodies, including the European Economic Community. BR inaugurated its part of the new station on 22 March 1982, when its old station was closed. The FR followed on 25 May, two days after the 150th anniversary of the act of incorporation.

Whilst great effort was being made to regain the Blaenau terminus, many changes had also taken place on the operating railway. The Harbour station headquarters was extended to give larger sales, catering and office space. Boston Lodge works had a new erecting shop and metal working facilities, enabling the building of a new Fairlie, *Earl of Merioneth*, which emerged in 1979, and six steel-bodied coaches. From 1970 the locomotives were converted to burn oil, eliminating the fire risk in the afforested areas of the line. Also at Boston Lodge a new carriage shop resulted in an improvement in the condition of the timber-bodied coaches. The former exchange yard at Minffordd was developed as a permanent way depot and road/rail interchange. A volunteers' hostel was provided here, following the successful conversion of the Penrhyn station buildings for the same purpose.

Between 1978 and 1982 the inclusive tours to Llyn Stwlan (the upper lake of the CEGB pumped storage scheme) and the slate caverns were operated via Tanygrisiau, buses completing the link. From 1982 the Blaenau interchange has been marketed with BR as the 'Festiniog Link' and has encouraged the promotion both circular tours and excursions using the Cambrian coast and the Conwy Valley lines as well as the FR. Since 1982 the Llyn Stwlan tour has been replaced by a Portmeirion connection, otherwise these arrangements still stand. The existence of the link enabled BR to extend its Red Star parcels service to Porthmadog and stations on the Lleyn Peninsular, making the FR the only private railway to carry parcels but a facility which only really came into its own when the National Eisteddfod was held in Porthmadog in 1987.

1986 was an eventful year which saw the railway celebrate its 150th anniversary. *Prince* went into traffic in a 1900s red livery; a gala weekend was held, with many attractions laid on for an appreciative audience; a 7¼in gauge railway operated at Tan-y-bwlch; de Winton vertical boilered locomotive *Chaloner* visited from Leighton Buzzard; BR Class 47 locomotive No 47645 was named *Robert F. Fairlie Locomotive Engineer 1831-1885* at Blaenau Ffestiniog; ex-Navy Planet diesel was restored for use on passenger trains and named *Conway Castle*; *Mountaineer* visited the Vale of Rheidol Railway; *Britomart* visited the Leighton Buzzard Narrow Gauge railway and a gravity slate train was operated from Blaenau Ffestiniog.

The return to Blaenau was achieved only at the expense of a large overdraft and it was found that further development was hampered by the need to service it. Solutions considered to improve the railway's capital base were in turn hampered by the Victorian statutes which governed it. Eventually a scheme was devised which calculated that there was a certain amount of prestige to be gained by issuing stock which had been authorised by Parliament in the last century, taking advantage of a small amount of unissued stock. It offered £1 shares with the proviso that each application for one share be accompanied by another for a £250 4% debenture, requiring a minimum £251 investment. The company's archivist, Michael Seymour, designed suitable certificates in a Victorian style and a lavish prospectus was printed. Launched in October 1987, over £¼million was raised.

The success of the 150th anniversary gala led to the operation of a similar event two years later, commemorating the 125th anniversary of steam traction; the highlight of this was the operation of the Welsh Highland Railway's *Russell*; the return to traffic of *Merddin Emrys*, newly overhauled and resplendent in Victorian outline and livery; and the operation of a mixed train in the traditional manner.

The railway's finances were not the only aspect requiring attention after 1982, for many parts of the line looked decidedly run down. Under the enthusiastic leadership of local resident, Eileen Clayton, the Parks & Gardens Department was established and a start was made to 'de-tattyfy' the line. Activities including clearance, planting, painting and refurbishing of buildings now regularly attract parties of over 100 volunteers, including paid staff off duty, the very young, the less young and many who are new to volunteering.

All this voluntary effort has been supported by the company, which has erected a platform canopy at Porthmadog and instituted a new regime, and livery, for carriage painting, whereby the whole fleet can be repainted every three years.

Attention has also been paid to the railway's heritage with the creation of a new museum in the Porthmadog Goods Shed, the turning out of *Prince* and *Merddin Emrys* (*Blanche* in Penrhyn colours is proposed) and several carriages in historic liveries. The operation of gravity and mixed trains comes into the Festiniog Railway Heritage Group's sphere of influence and has encouraged volunteers to restore several wagons and vans for use on special occasions. The last item of unrestored carriage stock, No 10 van, which was used in the early years of revival and then put to one side is now being put back into running order for use with the Vintage Train, the operation of which now being a regular operation.

The close of the 1980s brought further opportunities for development with the aid of outside funding although the results won't be seen until the 1990s. Known as INCA, for the scheme is to INcrease CApacity, the scheme will provide the following: three new carriages equipped with heating and public address; a 10-car carriage shed with maintenance facilities at Boston Lodge; a boiler and other major components for a new double Fairlie of traditional outline, allowing the *Earl of Merioneth's* superstructure to be retired (the availability of spare power bogies allows one to dream of the occasional operation of *three* double Fairlies on special occasions!); improvements to the buffet/bar at Harbour station; new toilets and improved cafe facilities at Tan-y-bwlch; permanent platform buildings and a carriage siding at Blaenau Ffestiniog; and a new diesel locomotive for exclusive use of the permanent way department.

Finally, there is an exciting scheme now in hand to provide the Festiniog Railway with a brand new locomotive by the year 2000. Called the 'Taliesin 2000' project, it will fund the building of a new 0-4-4T single Fairlie after the style of *Taliesin*, a G. P. Spooner designed locomotive, built in 1876 and withdrawn in 1932; for many the single Fairlie, a feature of the North Wales Narrow Gauge Railways and the Welsh Highland Railway as well as the FR, is a vital part of the line's heritage and something which should be restored. Over 220 will pay £60 a year for 12 years in support of their beliefs.

BIBLIOGRAPHY

Boyd, J. I. C.; *The Festiniog Railway*; Oakwood Press, 1975 (2 vols)

Festiniog Railway; *Ffestiniog Railway Traveller's Guide*; FR Co, 1989

'Taliesin'; *Festiniog Railway Locomotives*; AB Publishing, 1988

Below:
Double Fairlie *Earl of Merioneth* entered traffic in 1979 looking a product of its time. Ten years later it received attention, including the fitting of brass domes (ex *Merddin Emrys*) and a new paint scheme, to soften its lines as shown at Boston Lodge on 20 July 1989, shortly before it re-entered traffic. *Norman Gurley*

SPECIALS AND HEADBOARDS

Right:
The driver of the Talyllyn Railway's *Talyllyn* and the train guard wait for passengers, and the photographer, at Brynglas on 16 May 1987. The occasion was the locomotive's inaugural run following a lengthy overhaul; the event coincided with the launch of a book about it and its partner, *Dolgoch*, published by Ian Allan Ltd.

Centre right:
The inaugural run of the Festiniog Railway's 'Mountain Prince' service took place on 22 May 1987, with the veteran *Prince* as the train engine; it is seen on that occasion at Minffordd. The 'Mountain Prince' is the departure from Porthmadog which makes the first onwards connection with BR at Blaenau Ffestiniog. The red and white off-white carriage livery was originally developed for this service but is now applied to all except the oldest vintage stock. *Norman Gurley*

Below:
The Vale of Rheidol Railway's named train, 'The Welsh Dragon' is seen approaching Glanrafon on 21 August 1987. *D. Trevor Rowe*

Bottom left:
Bala Lake Railway signalman waits to take the single line token from *Holy War's* fireman at Llanuwchllyn during 'Teddy Bears' Weekend' in July 1989. Over 300 children accompanied by both teddy bears and adults travelled on the railway during the event. *Pete Briddon*

Bottom:
Talyllyn Railway No 3, *Sir Haydn,* in 'Thomas the Tank Engine' guise at Brynglas blockpost on 19 June 1983. *Tom Heavyside*

Right:
A selection of Vale of Rheidol headboards photographed in the loco shed store.

Centre right:
When the Welshpool & Llanfair Railway decided to run a 'Thomas the Tank Engine' event sponsorship was obtained from a building society to repaint Kerr Stuart 0-6-2T *Joan* blue, coincidentally the sponsors' corporate colour; the face was borrowed from the standard gauge Llangollen Railway. The venture proved to be the most successful ever run by the railway, with over 3,000 passengers being carried in two days. The locomotive is seen taking water at Welshpool on 8 July 1989. *Ralph Cartwright*

Bottom right:
For the Welshpool & Llanfair Railway's 25th anniversary it was possible to turn out one of the line's original locomotives, *The Countess,* seen here carrying a headboard at Raven Square on 1 October 1988.
Mike Haddon

THE WELSH NARROW GAUGE RAILWAYS

Talyllyn Railway

The 2ft 3in gauge Talyllyn Railway was opened in 1866. Including branches it was about 10 miles long and was designed to carry slate from the Bryn Eglwys quarries to a wharf adjacent to the Cambrian Railways' coast line at Towyn, in the old county of Merioneth. The railway company, established by an 1865 Act of Parliament, was a wholly owned subsidiary of the Aberdovey Slate Co, then recently formed to quarry the slate as a diversion from the depressed Manchester cotton trade, itself affected by the American civil war. It was originally proposed that the railway should terminate at Aberdovey, to take advantage of the harbour there, but construction of the coast line caused these plans to be modified. The railway is thought to have taken its name from the parish in which its traffic source and eastern terminus were situated; there is a lake bearing the same name three miles to the east of the station.

The line runs along the southern valley side of the Afon Fathew, opposite the mass of the Cader Idris range. The statutory Talyllyn Railway terminated at a spot some half a mile from the village whose name it bears; access to the quarries was by a branch line three quarters of a mile long, which made an end-on connection to the main line, and a series of inclines. On opening the minimum of equipment was provided, two locomotives, called *Talyllyn* and *Dolgoch*, four four-wheel carriages and a fleet of wagons. With the addition of a further carriage this stock sufficed for the Talyllyn Railway for over 80 years.

Wharf station, the coastal terminus, and a quarter of a mile from the standard gauge station, was for goods only, passengers being accommodated at Pendre, on the edge of the town. Pendre was also the location of the railway's depot and works. The main intermediate station became Dolgoch, where a locomotive water supply was laid on, and close to a fine three-arch viaduct, 62ft high and the line's only major structure, by which it crossed a ravine. The nearby falls are a continuing attraction for visitors. Until 1976 Abergynolwyn, 6½ miles from Wharf, was the inland passenger terminus. Goods traffic did reach the village by rail, however, as a branch off the mineral line, which included a cable-worked incline, served a goods shed, a sawmill, the Railway Hotel, as well as two terraces of houses.

For most of its existence the Talyllyn Railway was not overladen. Two return trips on six days a week were usually sufficient for the traffic offering, with most trains running mixed. Those few outsiders who came across the line soon decided that it had a charm all its own, especially when they found that the carriages only had doors on one side and that the brake van came with a fully equipped booking office!

The local MP, Sir Henry Haydn Jones, took over the quarries and the railway in 1911 and managed to keep the railway going until he died on 2 July 1950; his executors ran it until the end of that season. (The remaining working quarry was closed after the roof collapsed in 1946.) As the railway had been run on a shoestring for so many years it had become very run down; to many visitors it seemed that the track was only held together by the grass through which it ran! It seemed likely that the only future for the Talyllyn Railway could be closure.

In 1950 there were no preserved railways and the idea of amateurs involving themselves in railway operation was completely alien, both to enthusiasts and the public. Already, attempts to reopen the Festiniog Railway, closed in 1946, with the aid of volunteers had been unsuccessful.

However, following Sir Henry's death a meeting of enthusiasts was held in Birmingham to consider the matter of the Talyllyn. It was decided to form a preservation society and to approach the executors of the Haydn Jones estate. Following negotiations it was agreed to form a new company to hold the TR's shares in which both parties would be represented. In this way the first preservation society was formed and with its support the Talyllyn Railway reopened for business in 1951. L. T. C. Rolt, who had called the Birmingham meeting was appointed to manage the railway; his account of his experiences in 1951 and 1952, entitled *Railway Adventure*, is a classic of railway literature.

In addition to running the trains the new society undertook to search for new

Left:
No 2 *Dolgoch* at Nant Gwernol terminus with a special vintage train on 27 February 1982. The occasion was the 'Talyllyn Railway Pioneers of Preservation' residential course held at the Snowdonia National Park Study Centre at Plas Tan-y-Bwich.

equipment so that the 1866 stock could be repaired and augmented. Fortunately the nearby Corris Railway had been of the same gauge and much of its equipment, including two locomotives, was available for sale despite the three-year interval since the line had been closed; on the Talyllyn the locomotives became No 3 *Sir Haydn*, taking the name of the previous owner, and No 4 *Edward Thomas*, the line's manager for many years. Other material was recovered from greater distances; for example rail was obtained from Nuneaton and Crich – an exercise which was also to lead to the Tramway Museum Society being established at the latter place. A Barclay 0-4-0WT, now No 6 *Douglas*, was donated by a contractor following an appeal to industry for unwanted equipment. In some cases the revitalised railway was offered items which were of no use to it operationally – the Narrow Gauge Railway Museum was the result.

In the following years the track was made safe throughout and then, as finance allowed, relaid; the engines and rolling stock were repaired and rebuilt; and 'new' carriages, including an ex-Corris Railway and two ex-Glyn Valley Tramway bodies were obtained and rebuilt. An attractive fleet of bogie carriages was built to Talyllyn specifications and diesel locomotives obtained for engineering use, both new features for the railway. Wharf station was adapted for passengers and enlarged and Abergynolwyn was provided with a 'tea van' which commuted up and down the line until it was replaced by a permanent building in 1969.

A major task in early society days was the reconstruction of an embankment which had slipped, due to a blocked drain, near Dolgoch. To increase the line's operating capacity crossing loops were installed at Pendre, Brynglas and Quarry Siding.

The works area at Pendre has undergone great change over the years, with two carriage sheds, one of which has a paintshop area and an inspection pit, whilst the other is adjacent to a woodworking shop. In 1968 the cottage which was part of the complex, adjacent to the engine shed, became part of the shed, permitting four locomotives to be kept here. The rather primitive equipment which existed in the machine shop in 1950 has been replaced and augmented but the original screw-thread cutter has been retained.

From the beginning of Society operation there was a demand from some members for the mineral extension to be made fit for passenger trains. This would give an operating railway of 7½ miles with a terminus in the Nant Gwernol ravine. Work started on easing gradients and curvature in 1970 and the extension opened in 1976. In conjunction with the landowners a series of footpaths has been built around the new station for the benefit of passengers alighting there.

Above:
In 1982/3 No 3 *Sir Haydn* became the first Talyllyn locomotive to assume its 'Thomas the Tank Engine' character, *Sir Handel*, when the engine carried a face and was painted red. Afterwards its temporary name was removed whilst it continued operation in its new livery and is seen in this form at Nant Gwernol in 1985.

The depression of the 1980s had a considerable effect on the Talyllyn Railway, reducing the number of passengers carried for several successive years. To generate additional interest ex-Corris Railway No 3 was repainted and given a face to make it *Sir Handel* of the Skarloey Railway, the Rev Awdry character which was identified with this engine; it ran in this form in 1982 and 1983. The name and face were removed when, in April 1984, it was sent on tour with the Mid-Wales Development Board's exhibition train. (*Dolgoch* had undertaken a similar tour two years earlier.) The tour culminated in the exhibition train being present at Euston when a BR electric locomotive, No 86258, was named *Talyllyn* on 30 April. TR livery was restored to No 3 when the saddle tank was replaced in 1987.

Ex-Corris Railway No 4 was the next Talyllyn locomotive to receive paintshop attention when, in 1985, it was turned out in Great Western Railway 1930s livery, complete with 'brass' numberplates, as a tribute to its years in GWR ownership and also to commemorate that railway's 150th anniversary; two wagons and a van of Corris origin also received GWR livery at this time. No 4 featured in the Rev Awdry books as *Peter Sam* so in 1988 it appeared in service painted red with blue lining, with the name in yellow on the tanksides and carrying a smokebox face. (The Rev Awdry, incidentally, is a founder member and volunteer of the TRPS and he and his family have close links with the railway.)

The modernised facilities at Pendre works have generated a useful income for the railway over the years. In addition to jobbing tasks undertaken for local businesses and farmers several useful orders have been obtained from other railways, including the Teifi Valley and the Leadhills & Wanlockhead Railways.

The facilities at Pendre were of considerable value to the railway itself, of course, in the largest task to be undertaken there. In 1969 the railway obtained a 3ft gauge Barclay 0-4-2T from the *Bord Na Mona*, the Irish Peat Board. The locomotive, built in 1949, had seen little use and it was proposed to convert it to work on the TR. Some work was done and a name, *Irish Pete*, a pun on its background and originally a joke, and number, 7, allocated. The decline in traffic took the pressure off both the existing fleet and the need to complete No 7 so it spent most of the next two decades languishing in Pendre yard. In 1988 it was decided to resume work on the locomotive so that it could enter service in 1990, the year in which the railway celebrated its 125th anniversary and preservation society its 40th. At the society's 1989 AGM a new vote on No 7's name was conducted and the meeting decided to honour the society's founder by naming it *Tom Rolt* – a fitting tribute to the man and an eminent celebration for both society and railway. The locomotive is scheduled to enter service in 1991.

BIBLIOGRAPHY

Boyd, J. I. C.; *The Tal-y-llyn Railway*; Wild Swan, 1988

Johnson, Peter & Weaver, Rodney; *Great Preserved Locomotives 6 Talyllyn Railway No 1* Talyllyn & *No2* Dolgoch; Ian Allan Ltd, 1987

Rolt, L. T. C.; *Railway Adventure*,; Constable, 1953

LOCOMOTIVES

Right:
Barclay 0-4-0T *Dougal* was built for service at the Glasgow Corporation's Provan Gas Works in 1946. It has been based at the Welshpool & Llanfair Railway since 1969 and is regularly steamed on bank holidays and other special occasions. Here it comes under observation on 28 August 1989.

Centre right:
Talyllyn Railway Nos 1, *Talyllyn*, and 2, *Dolgoch*, were supplied to the railway by Fletcher, Jennings of Whitehaven and were the line's sole motive power until after the preservation society took over in 1951. On 16 May 1987 they are seen together at Tywyn Wharf.

Below:
The front ends of the Snowdon Mountain Railway locomotives are usually obscured by the carriages they propel but on 20 August 1986 No 3 *Wyddfa* stands outside Llanberis shed, waiting to go into service. The locomotive is one of three built in Switzerland in 1895.

Bottom left:
The double-heading of the day's last train allows the contrast in size and outline of the Welshpool & Llanfair Railway's Peacock 0-6-0T *The Countess* and the Stoke-on-Trent-built 0-6-2T *Joan* to be easily seen on 28 August 1989. The latter has regained its name but retains the blue livery of its 'Thomas the Tank Engine' identity; it has also had its balloon stack chimney replaced by one with a more traditional appearance.

Below left:
The Belgian-built 2-6-2T from Finland has proved to be major attraction at Llanfair Caereinion, especially as visitors are given access to the footplate. The name *Chernobog,* is not officially applied and is a Slavonic word meaning 'God of Evil'! The locomotive is now officially named *Orion.*

Bottom left:
The Welsh Highland Railway's *Karen* comes under youthful scrutiny at Pen-y-mount station on 30 April 1983. The locomotive was built for the Selukwe-Peak Light Railway of the Rhodesia Chrome Mines Ltd, in Northern Rhodesia, in 1942.

Above right:
One of the Festiniog's Railway's ex-Penrhyn Quarry Railway locomotives *Linda,* shunts restored slate wagons at Boston Lodge on 23 September 1988. Snowdon is the peak to the right.

Centre right:
Double Fairlie *Merddin Emrys* receives attention at Boston Lodge before being put away on 23 September 1988. Much of the work done on this locomotive in the 1960s and 70s was not carried out in a sympathetic manner. The generous support of a sponsor enabled the locomotive to be restored to pre-1946 outline and 1880s livery in time to celebrate the 125th anniversary of FR steam traction celebrated in 1988.

Right:
Pontsticill is the location for this view of the Brecon Mountain Railway's *Graf Schwerin-Löwitz* fitted with its tender on 29 August 1984.
Peter Groom

Below:
Vale of Rheidol Railway No 9 *Prince of Wales* puts up a good head of smoke as it crosses the Rheidol river bridge with the 10.30 departure from Aberystwyth on 19 June 1983. The Vista car is immediately behind the engine. *Mick Roberts*

Bottom right:
Beddgelert **and** ***Russell*** **in the Fairbourne Locomotive Works of the Fairbourne & Barmouth Railway.**

Right:
Boilers belonging to the Vale of Rheidol Railway No 9 *Prince of Wales* and a Baldwin Pacific receive attention at the Brecon Mountain Railway's Pant Works on 6 August 1989. The Rheidol locomotive's wheelsets are in the foreground.

Centre right:
The overhaul of the Llanberis Lake Railway's *Wild Aster* was approaching completion when photographed in the railway's Gilfach Ddu depot in August 1987.

Below right:
Aberystwyth engine shed on 14 August 1989, during the first season of private owership. The locomotives are No 8, No 7 and No 10; under BR ownership the steam locomotives were classified Class 98 and the diesel Class 98/1.

Below:
Privately-owned *Britomart* and a Festiniog Railway slate wagon in Boston Lodge Works for attention on 9 March 1986.

Bottom left:
On 18 August 1989 a view through the frames of the Snowdon Mountain Railway's No 6 *Padarn* reveals the locomotive's rarely seen drive pinion.

Bottom right:
At Pendre Works on 15 August 1988 the Talyllyn Railway's rebuilt Barclay 0-4-2T *Tom Rolt* (formerly *Irish Pete)* starts to look like a complete locomotive again.

Above:
Hunslet 0-4-0ST *Alan George* basks in the sun at Pontprenshitw on the Vale of Teifi Railway on 28 May 1989. *Mike Matthews*

Welshpool and Llanfair Light Railway

Attempts to build a railway to serve Llanfair Caereinion, a small town about eight miles to the west of Welshpool, were first made in 1864; it was 1903 before the 2ft 6in gauge railway eventually constructed was opened to traffic. Using powers obtained under the 1896 Light Railway legislation, the Welshpool & Llanfair Light Railway company was able to persuade its larger standard gauge neighbour, the Cambrian Railways, to build and operate its line, with construction work starting in May 1901.

The railway was distinguished by the route taken, from Welshpool's Cambrian station, through the town. It passed the cattle market, crossed the Shropshire Union Canal and the main street and then, straddling the Lledan Brook, the line ran between the backs of the houses before reaching the edge of the town and open country at Raven Square. This route shoe-horned the railway into the smallest space available, requiring the least amount of demolition.

Outside Welshpool the route mainly followed the Sylfaen Brook and the Afon Banwy, a tributary of the Afon Vrnwy, and included the Golfa bank, a gradient of 1 in 29 climbing for nearly a mile out of Welshpool; the six-arch Brynelin Viaduct, a fine stone structure which carries the line over a minor stream; and a bridge of the Banwy. It was this latter which nearly brought about the downfall of the preserved railway when it was severely damaged during a storm in 1964.

Unlike most of the other lines in this book which were always narrow gauge the Welshpool & Llanfair Light Railway was built to serve local agriculture; the only mineral traffic was to be incoming coal. To work its traffic the railway was provided with two Beyer Peacock 0-6-0T locomotives, named *The Earl* and *The Countess* after the Earl and Countess of Powis – the Earl had given land for the railway and was a director and shareholder. To carry it there were three bogie coaches and a fleet of goods vehicles, including brake, goods and cattle vans and open wagons. Timber traffic was catered for by the later provision of bolster wagons.

The railway was not to be the success its promoters had forecast and it was never to pay a dividend to its shareholders. Small operating surpluses were sometimes made but these went to pay interest on loans. The shortage of capital was such that a number of improvements were only undertaken with the aid of deferred payment arrangements made with the Cambrian Railways.

The '1922 Grouping of railways saw both the Cambrian and the Welshpool & Llanfair Railways being absorbed into the Great Western Railway in 1922. Everyone who had invested in the narrow gauge line lost money, the ordinary shareholders receiving only 4/11d (24½p) in the £1.

The takeover did not immediately affect W&L operations, except that from 1925 the GWR introduced a competing bus service between Llanfair Caereinion and Welshpool! The buses were intended to compete with other road services in the area but obviously attracted passengers from the three daily mixed trains because they were much quicker. A GWR report of 1926 recommending closure was not implemented but it did show that costs could be reduced if operations were restricted to a single shift, only possible if there were no passengers to deal with. At this time goods traffic was increasing but still not sufficiently to cover costs. It was inevitable, then, that the passenger service was withdrawn in 1931; from the same date only one gang, instead of two, was responsible for track maintenance. The coaches were put into store at Swindon Works and scrapped five years later.

During the closing years of the 1930s the railway settled into an operating routine which was to continue into the 1950s. There was one return working most days, leaving Welshpool at 11am and arriving back at about 4pm.

The Great Western did look after the locomotives, sending them, one at a time to Swindon for new boilers in 1929 and 1930; they returned home thoroughly 'Swindonised' in appearance, complete with copper capped chimneys and brass safety valve bonnets and number plates. In the first flush of ownership, the GWR provided two cattle vans to the Welshpool line and in 1924 four new bolster wagons were provided for the timber traffic.

Upon nationalisation in 1948 the line became a part of the Western Region and closure was again considered. The realisation that the considerable sum just spent, by the GWR in 1947, on repairing both engines would be wasted was enough to provide a stay of execution. This good fortune could not last, however, and the local authorities were advised of closure proposals in 1950. As with the passenger service, discussions could not prevent the inevitable. On 2 November 1956 the last goods train ran, followed by a passenger carrying special the next day, the participants travelling in brake vans and open wagons.

The Countess had already been removed to Oswestry Works and *The Earl* followed soon after closure. Arrangements were soon being made for the dispersal of stock and appointing a contractor to dismantle the line.

Since 1952 the Narrow Gauge Railway Society had been proposing preservation for the Welshpool & Llanfair Railway and had found that the British Transport Commission was prepared to sell the line for this purpose. By the time closure came the volunteer-supported Talyllyn and Festiniog Railways were demonstrating what was possible, although no-one

Left:

Leaving Sylfaen for Llanfair Caereinion, *The Countess* hauls a train of both Austrian and British-built stock from Sierra Leone on 28 August 1989.

Above:
The attractive bogie coaches being shunted by Sierre Leone No 85 at Welshpool on 24 May 1987 were Independence Day gifts from the British Government in 1961. The signal box is presently the only permanent building at the railway's eastern terminus but plans are being made for a station building which will provide the facilities expected by today's passengers.
Tom Heavyside

had yet bought a part of the state-owned system with the intention of running a public service on it.

The Welshpool & Llanfair Railway Preservation Society was established in 1956 and commenced negotiating with the Commission. One of the difficulties encountered was the Welshpool Town Council"s insistence that the town section be abandoned to allow road improvements to go ahead; the preserved railway would have to terminate at Raven Square, on the edge of the town.

By 1959 negotiations were sufficiently advanced that work had commenced on clearing the undergrowth which was taking over the track. When sufficient clearance had been achieved horses were hired to move the stock which the society was buying from British Railways.

The year 1960 saw the creation of the company which would be legally empowered to take over and run the railway; the society was wound up in 1961. The details of a lease were agreed with the Transport Commission and the procedures for obtaining the necessary Light Railway Transfer Order were pursued. Meanwhile, a search for rolling stock was under way: passenger vehicles were obtained from a naval dockyard on the Medway, fortunately of the same gauge, and an appeal for funds to purchase the steam engines was put in hand. The 'new' carriages were moved to Welshpool in 1961 and the locomotive appeal had progressed sufficiently for *The Earl* to move at the same time; *The Countess* was to follow on 6 October 1962, three days after the Light Railway Order had been made.

The first passenger train run by the new management at Christmas 1962 was also the first public service for 31 years. After a winter of continuous volunteer effort the official reopening took place on 6th April 1963. The trains which ran that day also traversed the line through the town; by August the Council had completed its purchase of this section and had closed it for good. Welshpool had lost something which made both town and railway unique. Some relics remain, including some mixed gauge track set in concrete in the standard gauge yard.

The first regular trains ran from Llanfair Caereinion to Castle Caereinion – a distance of 4¼ miles. From 1964 services were extended a further mile to Sylfaen. Concern about the state of the track on this section, coupled with the inability of the small track gang to cope with the additional workload, was the reason for once again terminating services at Castle Caereinion before the end of the season.

A major setback was suffered shortly afterwards, on 13 December, when the bridge over the Afon Banwy was severely damaged by debris carried by rising waters during a storm. Fortunately the railway was able to call upon military assistance to repair it, the reopening taking place in August 1965.

The search for additional locomotives and rolling stock continued. The articulated Bagnall 0-4-4-T, *Monarch*, was obtained from Bowaters at Sittingbourne, Kent, in 1966. A friendship struck up with the similarly gauged Zillertallbahn in the Austrian Tyrol resulted in four four-wheeled coaches being given to the railway in 1968. A Franco-Belge 0-8-0T locomotive, built in Franch to a German design for the German forces in 1944, was acquired from the Steiermarkische Landesbahnen, Austria, in 1969 and is now named *Sir Drefaldwyn* (Montgomeryshire). (In 1990 *Monarch*, out of use for several years, was deemed surplus to requirements and in 1991 was sold to two Festiniog Railway supporters.)

As the railway became established and consolidated its position it was able to devote time to extending again, running to Sylfaen from 1972. The purchase of the freehold from British Rail in 1974 provided the impetus to complete the line back to Welshpool. Work started in 1978, with volunteers working alongside Government-sponsored job creation employees. In addition to rebuilding the track a station was built at Raven Square, where it was necessary to divert Sylfaen brook to make enough room. Finance to undertake this project came in the form of loans from members and grants from the Wales Tourist Board and the Development Board for Rural Wales. The extension was opened on 18 July 1981.

Maintenance of both locomotives and rollling stock is a continuing process on any railway and the Welshpool & Llanfair is no exception. During the late 1980s considerable progress has been made in improving the conditions under which it is undertaken, especially with the construction and fitting out of a machine shop at Llanfair Caereinion. Progress has also been made with improving the railway's undercover storage, with the provision of a carriage shed built on the site of the former timber yard at Tanllan, just outside Llanfair Caereinion station.

In 1978 *The Earl* was withdrawn for overhaul but as there was little chance of the overhaul taking place for some time arrangements were made for the locomotive to go on display at the National Railway Museum the following year; in 1980 it participated in the 150th anniversary celebrations of the Rainhill Trials, travelling to Rainhill on the museum's Weltrol. It returned to York for display until 1984 and was later transferred to the Birmingham Railway Museum, going on display there in 1987. In 1989 it was further transferred to the Didcot Railway Centre. Meanwhile the other Beyer Peacock, *The Countess*, rarely used since 1970, was not being neglected, for an extensive overhaul was started in 1978 by a Manpower Services Commission team; the work was finished by the volunteers who succeeded in returning the locomotive to traffic in August 1986.

The railway's 80th anniversary celebrations in 1983 coincided with the introduction of the Mid-Wales Festival of Transport, now a regular event held each July. Its continued success has led the railway to consider other events as a way of increasing both publicity and patronage. A 'Thomas the Tank Engine' weekend held in 1989 attracted a record number of passengers for a two-day period, even surpassing some bank holiday weeks! Other events in the calendar include a Teddy Bear's Outing, an 'Epicurean Express' dining train, a Postman Pat weekend and Santa steam specials.

BIBLIOGRAPHY

Cartwright, Ralph & Russell, R. T.; *The Welshpool & Llanfair Light Railway*; David & Charles, 3rd edition 1989

Johnson, Peter; '85 years of the Welshpool & Llanfair Light Railway'; *Railway World*, October 1987

Top:
Skarloey Railway No 4 ***Peter Sam*** shunting carriages at Pendre whilst substituting for No 4 ***Edward Thomas*** on the Talyllyn Railway on 15 August 1989.

Above:
Prince shunting at Boston Lodge on 3 November 1986.

Left:
Maid Marian's driver keeps an eye on the photographer as he returns to Llanuwchllyn on a fine August day in 1987.

Top right:
Russell and ***Merddin Emrys*** at Harbour Station on 1 May 1988. *Peter Treolar*

Right:
Vale of Rheidol Railway No 8 in Cambrian Railways' livery approaches Glanrafon returning from Devil's Bridge in September 1988. *Andrew Banister*

30

CAMBRIAN

DIESELS

Right:
The Welshpool & Llanfair Railway has owned ex-Navy *Chattenden* since 1968 and uses it for both works trains and off-peak passenger services. It is a Baguley machine built in 1949 to an order from the Drewry Car Co. On 20 August 1989 it was found to be at Tanllan, just outside Llanfair Caereinion, where track was being laid into the railway's new stock shed.

Centre right:
***Conway Castle* passes Boston Lodge Works with one of the Festiniog Railway's experimental winter services to Rhiw Goch on 14 February 1990. The Planet locomotive was built for the Navy in 1958 and was acquired by the FR in 1981. In 1985/6 a major overhaul saw it regauged from 2ft 6in, equipped for push-pull working and fitted with a Boston Lodge-designed superstructure. Car 121, behind the locomotive, is carrying the second version of the railway's green-and-off-white push-pull livery to be applied to it.**

Bottom left:
Snowdon Mountain Railway No 9 *Ninian* was built to the railway's own design and specification by Hunslet in 1987. It was photographed at the summit on 18 August 1989.

Bottom right:
In 1983 Baguley-Drewry built six locomotives to an order from the Instituto Nacional do Acucar in Mozambique. The order was subsequently cancelled and the locomotives, and the components for four more, remained unsold when Baguley-Drewry went into liquidation. In 1985 the Festiniog Railway received 3767 on loan so that it could be demonstrated to potential purchasers; limited use was made of it as the size of its cab precluded it passing through Rhiw Plas bridge or Garnedd Tunnel. Three of the locomotives, including 3767, were sold to the Bredgar & Wormshill Light Railway in Kent and it was removed from the FR in 1986. Part exchanging the Peckett, unused since acquisition in 1957, the FR bought 3767 for Permanent Way Department use two years later and the photograph shows it at Boston Lodge in August 1989 whilst its development was being assessed. Financed by the InCa scheme, the cab has been re-shaped to fit the FR's loading gauge and the locomotive has been named, bilingually, *Harlech Castle*.

Above:
The works plate fitted to Vale of Rheidol Railway No 10. When the Brecon Mountain Railway took over the locomotive as part of their line purchase in 1989 they found it had hardly been used.

Right:
The four sets of Baguley-Drewry locomotive components from the cancelled export orders were bought by the Brecon Mountain Railway who, in 1987, built themselves a locomotive, shown outside Pant Works, with the frames of Vale of Rheidol No 9, on 6 August 1989.

Below:
Fairbourne & Barmouth Railway bogie diesel hydraulic *Lillian Walter* started life as the 15in gauge *Sylvia*, built by G & S Light Engineering in 1961. It re-entered service with an American outline body and a Mercedes engine in 1985 and the following year was regauged to 12¼in. August 1986.

Above:
Diesel No 10 *Yeti* crosses the Afon Hwlch viaduct on the Snowdon Mountain Railway as it returns to Llanberis on 18 August 1989.

Vale of Rheidol Light Railway

At the end of the 19th century the Rheidol valley, near Aberystwyth, was a hive of industrial activity. The various metal mines in the area were in increasing need of improved communications between them and the port, and growing resort, of Aberystwyth. The Manchester & Milford Railway had obtained powers to build a standard gauge line to Devil's Bridge, at the head of the valley, in 1861 but the scheme did not come to fruition.

The miners, and others, had to continue using the poor quality road network with pack animals until 1902, when the 1ft 11½in gauge Vale of Rheidol Light Railway was opened to traffic. In the 11¾ miles from Aberystwyth to Devil's Bridge the line climbs 680ft, 480ft of this being achieved on a gradient of 1 in 50 over the last four miles from Aberffrwd. As might be imagined the route sticks closely to the contours in a way that no standard gauge line ever could, demonstrating the constructional value of building narrow gauge lines in mountainous districts.

To build the railway the contractor used equipment, including a 2-4-0T Bagnall locomotive from the nearby, and then recently closed, Plynlimon & Hafan Railway. The Vale of Rheidol took over the engine and wagons when construction was finished, naming the engine *Rheidol*. It was scrapped by the GWR in 1924 but some of the Plynlimon & Hafan wagons continue in permanent way department use.

To work alongside the Bagnall the company purchased two 2-6-2T locomotives built to its requirements by Davies & Metcalfe of Romiley. They were the only ones ever built by this company.

The new railway traded with a modicum of success, small profits being made each year. Indeed, in 1912 traffic was so heavy that an engine, *Palmerston*, had to be borrowed from the Festiniog Railway. In 1910 control of the line passed to nominee directors of the Cambrian Railways and in 1913 that company took over the narrow gauge company.

The new management did not appear to devote the same attention to the Rheidol line as did its original owners and the railway entered a period of gradual decline. It was still necessary, however, to borrow the FR's *Palmerston* in 1913 and 1914, an arrangement only terminated by the outbreak of war.

Above:
On the Vale of Rheidol Railway the 1980s were mostly noticeable for the attention paid to stock liveries. Here No 7 is seen in BR lined brunswick green on 19 June 1983, prior to the recommissioning ceremony which introduced it to the public on that date.

Mining in the Rheidol valley had been in decline since the railway opened and after World War 1 ended several of the mines closed, reducing still further the small amount of mineral traffic carried by rail. Passenger traffic did increase though, especially in the summer months and it was necessary to borrow *Palmerston* again in 1921 and 1922.

The Cambrian was then itself absorbed, by the Great Western Railway, in 1923 and a review of services was undertaken. Two locomotives were built at Swindon to designs based on those of Davies & Metcalfe in 1923 and in 1924 a third was erected using the spare parts made the previous year for the first two! For many years it was said, by the GWR originally, that the third Swindon-built locomotive, now No 9, was a rebuild of one of the Davies & Metcalfe machines; extensive research carried out by historian C. C. Green and published in his book on the VofR (see bibliography for details) shows that this was not the case. The reason for the deception was to fool the GWR's accountants as construction of three new locomotives would not have been sanctioned although they were considered necessary. The Bagnall and one of the Davies & Metcalfe locomotives were scrapped in 1924, the second D&M locomotive meeting the same fate in 1935.

Further activities under GWR ownership included closing the harbour branch, providing four open carriages in 1923, moving the Aberystwyth terminus to a site alongside the main line station in 1924 and replacing the original carriages in 1938. In 1923 two cattle vans were supplied but so little use was found for them, it being said that there was no cattle in the valley, that in 1937 they were regauged and sent to the Welshpool & Llanfair Light Railway. By the end of the 1920s the goods traffic had gone completely and in 1931 the winter service was withdrawn. The Vale of Rheidol Light Railway had become solely a tourist line.

In 1939 the line was closed for the duration of World War 2. Services were resumed in 1945, a prompt restoration being aided by the occasional maintenance attention given to the stock and track during the closure. Upon nationalisation in 1948 two of the engines and the carriages were turned out in the new owners' livery although the third engine, No 7, was to continue to proclaim allegiance to the GWR until 1954. The 1924-built locomotive, still carrying the number, 1213, of its predecessor became No 9 in sequence with the 1923-built machines, Nos 7 and 8, during the 1948/9 winter.

In 1956, probably spurred on by the success of the Talyllyn and Festiniog revivals, BR decided to make an effort to promote the line, naming the engines and painting the coaching stock in Western Region chocolate and cream livery. The resulting publicity did increase traffic and a year later the locomotives were turned out in lined brunswick green, the current express locomotive livery.

In the years following the 1963 transfer of the VofR from the Western Region to the London Midland Region the line assumed an air of neglect. The stations became overgrown; the crossing loops, at Capel Bangor and Aberfffrwd, were removed; internal communications were abandoned in favour of a GPO (now British Telecom) leased line; and, following a period when the carriages were painted dark green with a strange 'VofR' logo, the carriages and locomotives were painted a rather dreary unlined rail blue.

Things began to improve a little in 1968 when the Aberystwyth terminus was abandoned and the line diverted to a new location in the former Carmarthen line bay of the main line station. At the same time the stock was found a new home in the standard gauge locomotive shed.

The year 1970 saw the establishment, with BR support, of a supporters' association; this had the objective of promoting the line in a similar manner to the societies which support most preserved railways. The Association's effort, and increased publicity given to what had become BR's only steam operation, paid off.

The increase in passengers carried encouraged BR to invest in locomotive overhauls, carried out at Swindon, and extensive carriage repairs, carried out on a disused platform at Shrewsbury station by craftsmen from Chester Depot. By the time the programme was completed, by the late 1970s, both motive power and carriages were sporting black and white lining which brightened up the rail blue considerably.

Restrictions on operation during the high fire risk period of the dry summer of 1976 led to oil firing being adopted, No 7 being converted first. Both materials and and advice were obtained from Boston Lodge, the Ffestiniog Railway's fleet having already been converted for similar reasons. The work was carried out in the shed at Aberystwyth, the first of a number of major tasks which have been undertaken there.

Efforts to maintain the railway's attractiveness to both public and enthusiast continued into the 1980s, with a Cambrian lines 'rail week' being held at Aberystwyth in May most years from 1981. In addition to excursions being operated from England, exhibits and stalls were set up on the platform and the narrow gauge line was a centre of attraction in its own right, with extra trains being run.

Attention was also paid to liveries, with No 8 being given a simple GWR finish, as an experiment, in 1981; this was replaced by Cambrian Railways 'invisible green' (black in most lights) livery in 1986. No 9 had the original VofR company colours applied in 1982 and No 7 was BR lined brunswick green the following year. Also in 1983 carriages were turned out in Great Western livery. But for some vehicles the changes were rather more than cosmetic. The brake thirds were given a first class compartment, complete with upholstered seating, and one of the open cars, No 4999, was rebuilt, at Chester, with glazing and modified upholstered seating, as a Vista car. Most of these works were undertaken with the aid of external sponsorship.

A new item of self-propelled motive power was delivered on 29th October 1985, in the form of a Permaquip personal carrier. It replaced a 1963-supplied Wickham trolley which had been out of use since 1984 and was allocated to the permanent way department. Two years later another item of narrow gauge internal combustion motive power was delivered to Aberystwyth. This was an 0-6-0 diesel hydraulic locomotive built by the Brecon Mountain Railway using parts made by Baguley-Drewry for a cancelled export order. Given the number, 10, it was paid for by a grant from the Mid Wales Development Board. It was intended that No 10 should be used for out of season work and off-peak, three-car maximum, passenger trains.

Earlier, unwelcome media attention was given to the Vale of Rheidol Railway when on 26 May 1986, a busy bank holiday, the last two cars of the 16.00 from Devil's Bridge, hauled by No 7, derailed just below Nantyronen. Positive news was generated that year, however, when two carriages, Nos 4146 and 4734, were named *Lowri* and *Myfanwy* respectively on 3 August; when a Class 37, No 37426, was named *Y Lein Fach/Vale of Rheidol*; when the Festiniog Railway's Alco 2-6-2T *Mountaineer* was the star turn at an Autumn Gala weekend; and when it was announced that £100,000 was to be spent on track work.

The future of the line under British Rail ownership was becoming increasingly unclear, however, and rumours that the line was to be sold abounded. Eventually, in October 1987, it was confirmed that the line was for sale; the announcement was made at the same time as one concerning the 1987 passenger figures – the best for seven years and up 38% on the previous year!

The 1988 season's publicity was branded 'last year in British Rail ownership' although tender documents were not issued until 6 June, with a closing date of 19 July. To mark the end of an era an enthusiasts day, advertised as '20 years on – the real end of British Rail steam' was held on 5 November when an intensive service was operated and a series of santa trains were run on two weekends in December.

In August it was announced that the Brecon Mountain Railway had emerged as the preferred bidder with Government approval being given to the sale, for £306,500, by means of a Light Railway Order effective from 30 March 1989.

The new owners commenced operations on 21 May, with a service the same as that operated by BR. For the 44,000 passengers carried in 1989 the only outward change was the transfer of Aberystwyth ticket sales to a kiosk placed on the narrow gauge platform. At the same time the inquisitive enthusiast soon found that locomotive No 9 had been transferred to BMR headquarters in South Wales for a major overhaul. By the spring of 1990 two carriages had also made the journey southwards.

It will take some time for the new owners' plans for the Vale of Rheidol Railway to come to fruition; there is much to be done before the line can be considered to be in sound condition – much rail dates from 1902, many sleepers were installed in the 1920s, hedges are badly overgrown, the public expects shelter and entertainment while waiting for their trains and more than hard benches to sit on. Given time, and continued public support whilst the work is carried out, the Rheidol line will be in the forefront of those described in this book.

BIBLIOGRAPHY

Green, C. C.; *An Illustrated History of The Vale of Rheidol Light Railway*; Wild Swan, 1986

Below:
No 8 *(Llewelyn)* has now carried two historic liveries, the second being shown here at Nantyronen on 5 November 1988. Except for the Christmas services, this was the penultimate Rheidol train operated by British Rail.
Mike Jones

Above:
With Tal Fechan reservoir in the background the Brecon Mountain Railway's *Graf Schwerin-Löwitz* approaches Pontsticill on 6 August 1989.

Left:
Thomas Bach, traditionally known as *Wild Aster*, emerges from the Vivian Arch on the Llanberis Lake Railway on 16 July 1988.
Horace Gamble

Above right:
The Countess and *Joan*, the latter in 'Thomas the Tank Engine' livery, raising steam at Llanfair Caereinion on 28 August 1989.

Right.
With Barmouth Bridge in the background, the replica Lynton & Barnstaple locomotive *Yeo*, was caught by the camera in brilliant sunshine on 1 October 1989. *Mike Matthews*

No 2

SOUTHERN
759

GAUGE CONVERSION CONTRASTS

Right:
The Carmarthen Bay at Aberystwyth: in September 1962, showing the now demolished canopy. *Andrew Muckley*

Below right:
And on 5 May 1986, occupied by the Vale of Rheidol Railway. (No 8 was about to depart with its first train since being repainted in Cambrian Railways livery). *Hugh Ballantyne*

Bottom left:
Standard gauge trains crossing at Llanuwchllyn in the early 1960s, looking westwards.
Andrew Muckley

Bottom:
Llanuwchllyn as developed by the Bala Lake Railway.

Facing page, top left:
Henllan looking towards Newcastle Emlyn on 25 September 1971. *Andrew Muckley*

Top right:
The 4ft gauge Padarn Railway terminus, complete with transporter wagons, at Gilfach Ddu, Llanberis, on 25 June 1956. *R. E. Vincent*

Centre:
The Penrhyn Quarry Railway's 0-4-0T *Blanche,* built to 1ft 10¾in gauge by Hunslet in 1893, at Port Penrhyn on 28 May 1954. The bucket contained sand. *Ian Allan Library*

Bottom right:
***Blanche* on the Festiniog Railway in 1988, regauged to 1ft 11½in following purchase in 1963. In addition to the tender cab, the locomotive now has a leading pony truck, piston valve cylinders and superheating.**

THIRD

BLANCHE

LLANUWCHLLYN

Above:
Cnicht and *Britomart* on the Welsh Highland Railway in May 1988.

Snowdon Mountain Railway

The Snowdon Mountain Railway is a Swiss mountain railway transposed to North Wales. The mountain is the highest peak in England and Wales, the railway on it being the first of those here described to be built exclusively for its current traffic – tourists.

It was not the first railway to be proposed for Snowdon, that being put forward in 1871. If built it would have terminated 500ft from the 3,559ft summit. Further schemes were promoted in 1874 and 1881, the latter being a line from Llanberis to Rhyd Ddu, a station on the North Wales Narrow Gauge Railway, via the summit. These projects failed because the landowner refused to release any of his property to enable them to go ahead.

The climate had changed by 1894 when the Snowdon, Mountain Tramroad & Hotels Co Ltd was registered as a public company with the tacit support of the landowner. The company was empowered to build a railway up the mountain from Llanberis, with a hotel at the summit, and to take over the Royal Victoria Hotel at Llanberis. The railway was to be a rack line built to the 'Abt' principle and construction started in December 1894. Legal powers were not required because all the land required was in private ownership.

Whilst carriages, wagons and track materials were ordered from English firms, three steam locomotives were ordered from Switzerland, the centre of mountain railway expertise. The use of steam was considered to be provisional as electricity was actively considered during construction. The finished railway was 800mm (2ft 7½in) gauge and 4½miles long with a ruling gradient of 1 in 5½.

The official opening took place on Easter Monday, 1896. Despite the successful operation of numerous works and test trains, disaster struck. The locomotive of the first public down train left the rails just above Clogwyn and fell down the mountainside, damaging the telegraph as it did so. The breaking telegraph cables short circuited, causing the telegraph bell at the summit to ring. The crew of the train waiting there took this as the signal to leave. They did so and ran into the carriages of the first train. A man who had jumped off the first train fell under it and later died of his injuries. The passengers of both trains had to walk back to Llanberis. While investigations into the cause of the accident got under way the company decided to suspend operations.

It was eventually decided that some slight settlement of the track was probably the cause of the locomotive climbing the rack. To prevent a repetition flanged guard rails were fitted to the rack, the modification being completed before the railway was reopened to the public on 19 April 1897.

By this time the company owned five locomotives although the remains of one of them was in an irrecoverable location on the mountainside. They were built by Schweierische Locomotiv-und Maschinenfabrik at Winterthur, Switzerland, and had the appearance of being 0-4-2Ts with inclined boilers. The wheel arrangement is misleading, though, because the 'driving' wheels are actually loose on their axles, the drive being transmitted through pinions mounted on the leading axles. The inclination of the boiler prevents the firebox being damaged when the locomotive is on the gradient, a standard practice on mountain railways.

The coaches were built in England to Swiss specifications. One of the bogies is fitted with a brake pinion which engages if the speed exceeds a preset limit. Its efficiency was demonstrated on 6 April 1896. After the accident a second, independent, brake was also fitted. Originally the locomotives were to propel two carriages up the line but since the reopening a single carriage has been the norm; there is no coupling between carriage and locomotive.

Following the reopening considerable efforts was put into publicise the railway; the response showed that the accident was no deterrent. The additional safety precautions taken were probably a great morale booster.

The debentures issued to finance construction became due for repayment in 1915, in the midst of the First World War. Although the railway continued operating the receipts were on the low

Below:
A traditional scene on the Snowdon Mountain Railway, as a steam locomotive propels a carriage past the locomotive shed. The siding in the foreground is the first new one to be laid since the line opened and was funded by the 1985 share issue; explosives were needed to prepare the ground for it during the 1985/6 winter.

side so an agreement was obtained which delayed repayment for five years, with extra interest being paid. By the 1920s the railway was sufficiently well-off to buy additional rolling stock; Switzerland was the source of three new locomotives, up-to-date versions of the early ones, and four new carriages.

The company was restructured in 1928 and was renamed the Snowdon Mountain Railway Ltd. Since 1923 the Board of Directors has always included at least one member of the Davies family, Pwllheli-based solicitors who also had interests in the Festiniog and Welsh Highland Railways at that time.

Unlike many other organisations, the SMR made profits during the 1930s and entered World War 2 in good order. During the war the summit buildings were taken over by different sections of the armed forces and many trains were operated On His Majesty's Service. The operation, or not, of a passenger service appeared to depend upon the whims of the current tenants of the summit but no trains were run for the public from the end of 1943.

After the cessation of hostilities public services were not resumed until 1946. Lack of attention during the war meant that considerable effort was required to overcome the arrears of maintenance. The coaches were rebuilt with new fully enclosed bodies, one being dealt with at Llanberis each year from` 1951. From 1956 a programme of locomotive overhauls was commenced, the four eldest being sent, one each year from 1958, to Hunslet in Leeds for the work to be carried out; work on the others, including firebox replacement, was undertaken to Llanberis and this is the current practice. The track was relaid throughout from 1967. Mortgages were taken out to finance some of this work but, ultimately, it was only possible because the public still wanted to travel to the summit of Mount Snowdon by train.

Servicing the summit was, and is, an expensive business; all supplies, including water, and waste have to be transported by train. The situation is aggravated as many more people use the facilities than go there by train. Many changes have taken place at the summit over the years; the buildings were replaced completely in 1936; extensive alterations carried out in 1954 and 1965 and a licensed bar was opened in 1968. But by the 1980s the buildings had become decidedly shabby, the threat of compulsory purchase in 1972 deterring the company from spending money on them. The impasse was resolved in 1983 when ownership of the company's property passed to the Gwynedd County Council. The new owners started an extensive renovation scheme and leased the hotel (only staff had stayed there since the war) back to the company – who else would have sufficient experience to run it?

From the average passenger's point of view one of the less pleasing aspects of the Snowdon experience was the amount of time spent queueing to secure a seat on a train, especially in high summer. From 1984 a token scheme was introduced whereby tokens were issued for a particular train as would-be passengers arrive at Llanberis. Only those with correctly numbered tokens need to queue when the booking office opens for their train.

In January 1985 the Snowdon Mountain Railway plc launched a very successful share issue which was fully subscribed in eight days. Some of the new capital was used to finance the construction of two Hunslet diesel locomotives, identified as No 9 *Ninian* and No 10 *Yeti*, which were delivered the following year. Intended to complement the steam fleet, not to usurp it, they have proved to be very successful. In addition to enabling the company to run more trains it is very likely that their reduced operating costs, when compared to those of the steam locomotives, will enable the company to extend its operating season when prospective passengers offering are not sufficient to make a steam service economic. To supplement the diesels a new carriage was delivered from East Lancs Coach Builders Ltd in 1988. In designing the vehicle to SMR specifications full advantage has been taken of both the railway's loading gauge and modern coach building techniques; features include a centre gangway and wheelchair capability.

Above:
Another product of the 1985 share issue is the diesel fleet; No 10 *Yeti* is seen descending from Clogwyn on 18 August 1989, shortly after the carriage with it entered service.

To supplement its interests in tourism the railway company established Snowdon Mountain Aviation Ltd to operate Caernarfon Airport in 1986, flying pleasure trips in a Cessna and a de Havilland Rapide; an aviation museum was opened there in 1989.

The Snowdon Mountain Railway is in the unfortunate position of actually turning away more people than it carries; it is clear that it is well placed to develop schemes to increase its capacity during the 1990s and can look forward to its centenary with confidence.

BIBLIOGRAPHY

Boyd, J. I. C.; *Narrow Gauge Railways in North Caernarvonshire Vol 1 West*; Oakwood Press, 1981

Johnson, Peter; 'By Rack to the Abode of Eagles – The Snowdon Mountain Railway'; *Railway World*, March 1989

Turner, Keith; *The Snowdon Mountain Railway*; David & Charles, 1973

Right:
The Welsh Highland Railway's *Russell*, with the Festiniog Railway's vintage train at Minffordd during the 1988 FR Steam 125 gala.

Centre right:
***Livingston Thompson* was, in 1886, the second Fairlie to be built at the Festiniog Railway's Boston Lodge Works. Withdrawn from service in 1971 it was cosmetically restored to 1910 condition in 1988 and sent on loan to the National Railway Museum where it was photographed on 18 May 1989. The locomotive was included in the 'National Railway Museum on Tour' display at Swindon in 1990.**

Bottom left:
The first Garratt articulated locomotive was built by Beyer, Peacock in 1909 for service in Tasmania. It was returned to the makers for preservation in 1947 and sold to the Festiniog Railway when Beyer, Peacock closed down in 1966. Following a period on display in the open at Harbour Station it was put into store until 1976 when it was sent to the National Railway Museum on long term loan. Restored in grey photographic livery it is seen displayed on the Museum's Weltrol wagon on 18 May 1989. This locomotive was also included in the Swindon exhibition.

Bottom right:
De Winton 0-4-0VB *Chaloner* and *Britomart* at Minffordd during the Festiniog Railway 150th anniversary celebrations on 20 April 1986. Both locomotives were built, in 1877 and 1899 respectively, for the Pen-yr-orsedd slate quarries in Nantile Vale. *Chaloner* is normally resident at the Leighton Buzzard Narow Gauge Railway.
Hugh Ballantyne

Above:
In exchange for the loan of *Chaloner, Britomart* visited Leighton Buzzard for several weeks and is seen at Stonehenge on 21 September 1988.

Right:
The Festiniog Railway's *Mountaineer* visited the Vale of Rheidol Railway in 1986 and is seen near Rheidol Falls Halt on 13 September.
Hugh Ballantyne

Bottom right:
***Britomart* was a visitor to the Bala Lake Railway's summer gala in 1989 and is seen at Llanuwchllyn on 25 June.** *Mike Haddon*

Llanberis Lake Railway

The Snowdon Mountain Railway is not the only narrow gauge railway with a home in Llanberis. The other line to be found there, the Llanberis Lake Railway, is in some respects a newcomer, but is founded on a route far older than that of the Snowdon line.

The route is that of the Padarn Railway, the common name given to the Dinorwic Quarry Railway. This was a 4ft gauge line brought into use in 1843 to replace an earlier tramway and was built to convey the output of the Dinorwic slate quarries, just outside Llanberis, to the quarry-owned Port Dinorwic, on the Menai Straits,. Horse-worked at first, locomotives were introduced in 1848 when two 0-4-0s named *Fire Queen* and *Jenny Lind* were obtained from A. Horlock's Northfleet Ironworks in Kent. In 1882 and 1886 they were replaced by Hunslet 0-6-0Ts named *Dinorwic* and *Pandora*, later renamed *Amalthaea*. A third Hunslet, *Velinheli* was added to the fleet in 1895. Of the Horlock engines, *Jenny Lind* was scrapped but *Fire Queen* was put away in a shed at Gilfach Ddu where it remained until 1969; rumour has it that the owner's daughter had an affection for this locomotive! It is now on display at the Penrhyn Castle Museum, Bangor. The Hunslets were scrapped in 1963, after the line was dismantled by a scrap merchant.

Padarn Railway operations were unsual in that the 1ft 10¾in gauge quarry wagons were carried, loaded, to the port on 4ft gauge transporter wagons. No public service was offered but from 1895 until 1947 trains were run for the benefit of the quarrymen, using purpose built rolling stock paid for by the men themselves! (They also paid for *Velinheli* although they did not have exclusive use of it!!)

In the quarries the different levels were served by 1ft 10¾in gauge systems. Locomotives were used where the length of run or the amount of work warranted it; otherwise manpower was used. Most of the tracks at Port Dinorwic were of this gauge and engines were allocated there for shunting purposes. From 1870, 29 steam engines and a lesser number of internal combustion locomotives were used at varying times. The quarry workshops at Gilfach Ddu carried out running repairs to the locomotives; major repairs and rebuilds were carried out at the port by the Port Dinorwic Dry Dock Company, a linked organisation which was probably capable of building new locomotives had it been called upon to do so; Snowdon Mountain Railway locomotives were sent there for overhauls until the 1950s.

The Padarn Railway remained in use until October 1961. The quarries followed in 1969 when the company went into liquidation. Whilst much of the Padarn Railway's equipment was scrapped this was not to be the fate of the quarry equipment.

Proposals for using part of the Padarn Railway trackbed for a tourist railway were first made after the tracklifting had been completed in 1963. One scheme suggested linking the Padarn to the BR Llanberis branch (which was closed in 1964) to create a circular route around Llyn Padarn. The County Council's decision to use the BR trackbed for road improvements concentrated the minds of the proponents of a Llannberis tourist railway onto the two miles of Padarn Railway alongside the lake. No progress had been made when the quarries closed.

In December 1969 the liquidator organised a sale to dispose of the remaining assets. At the auction it was announced that the Caernarfonshire County Council had obtained a preservation order for the quarry workshops at Gilfach Ddu and much of the equipment therein. Subsequently the council was to buy all the buildings and land not directly associated with the quarries.

Below:
A lake railway train passes boat-borne observers as it leaves Gilfach Ddu on 18 August 1989.

The auction was to be a catalyst; with support from the council and the Wales Tourist Board the Llanberis Lake Railway was formed; later the company was reregistered in Welsh, becoming Rheilffordd Llyn Padarn. Three quarry engines and other material were bought for the new line at the sale. The quarries were to be the source of some of the rail required; it could not be laid until the undergrowth had been cleared from the trackbed – shades of the early Festiniog Railway revival!

The new railway company decided to abandon the quarry track gauge of 1ft 10¾in in favour of the more common 1ft 11½in. Whilst the clearance and tracklaying were carried out locomotive overhauls and carriage building took place.

By 19 July 1971 track laid as far as Cei Llydan, halfway along the lake, was fit for use and a passenger service was started. During the following winter the remaining track was laid to the terminus at Penllyn. The run-round loop at Ceim Llllydan was resited some 50yd closer to Gilfach Ddu and became a passing place; passengers are encouraged to leave the trains to use the picnic area here. Because of site and access restrictions passengers cannot board or alight at Penllyn. The railway company leased the engine shed in the old quarry workshop from the county council and it is here that the locomotives are stabled. Maintenance and overhauls are carried out here also. The carriage stock is kept in the former Padarn Railway workmen's train shed.

In 1971 the trains were hauled by one of the quarry engines, *Dollbadarn*, a 1922 Hunslet bought at the auction. The second, *Red Damsel*, dates from 1889 and was renamed *Elidir* before entering service in 1972. The third, *Wild Aster* of 1904, went into service in 1988, painted blue and named *Thomas Bach (Little Thomas*, a reference to the 'Thomas the Tank Engine' television series so widely appreciated by children). The railway's other locomotives are painted maroon. A further steam engine, 1984-built Henschel 0-4-0WT *Helen Kathryn* which was at the Bala Lake Railway between 1973 and 1975, and several diesels are also owned by the company.

During the remainder of the 1970s the railway consolidated its position as a new member of the 'Great Little Trains' marketing group. The effect of the varying numbers of tourists visiting the area each year was eased by the construction of the Dinorwic Power Station. This is a pumped storage scheme with its generating equipment situated in enormous caverns excavated under Elidir, the mountain on which the Dinorwic quarries were situated. The power generated is carried away by underground cables buried alongside the railway. The only means of access for the necessary excavating and cable laying? The Llanberis Lake Railway! The power station contracts started in 1975 and were to generate a considerable amount of work for a railway designed for tourists and were responsible for the entire main line being relaid with new rail. There is continuing evidence of power station activity in the form of a pumping station which was built between Cei Llydan and Penllyn, complete with a siding so that heavy machinery can be moved in and out by rail.

The development of the council-owned land around Gilfach Ddu as the Padarn Country Park explains the occasional reference to the Padarn Lake Railway. It has also had a positive effect on the railway; its station contains the only refreshment counter in the area.

The Llanberis Lake Railway may have no plans for great extensions in the future, but if it continues as it has in the past, quietly consolidating its position and improving facilities for passengers, then it is certainly a line with a bright future. That this view is shared by others is demonstrated by the takeover bid which the railway fought off in 1985!

BIBLIOGRAPHY

Boyd, J. I. C.; *Narrow Gauge Railways in North Caernnarvonshire Vol 3 The Dinorwic Quarry and Railways, Great Orme Tramway and other rail systems*; Oakwood Press, 1986

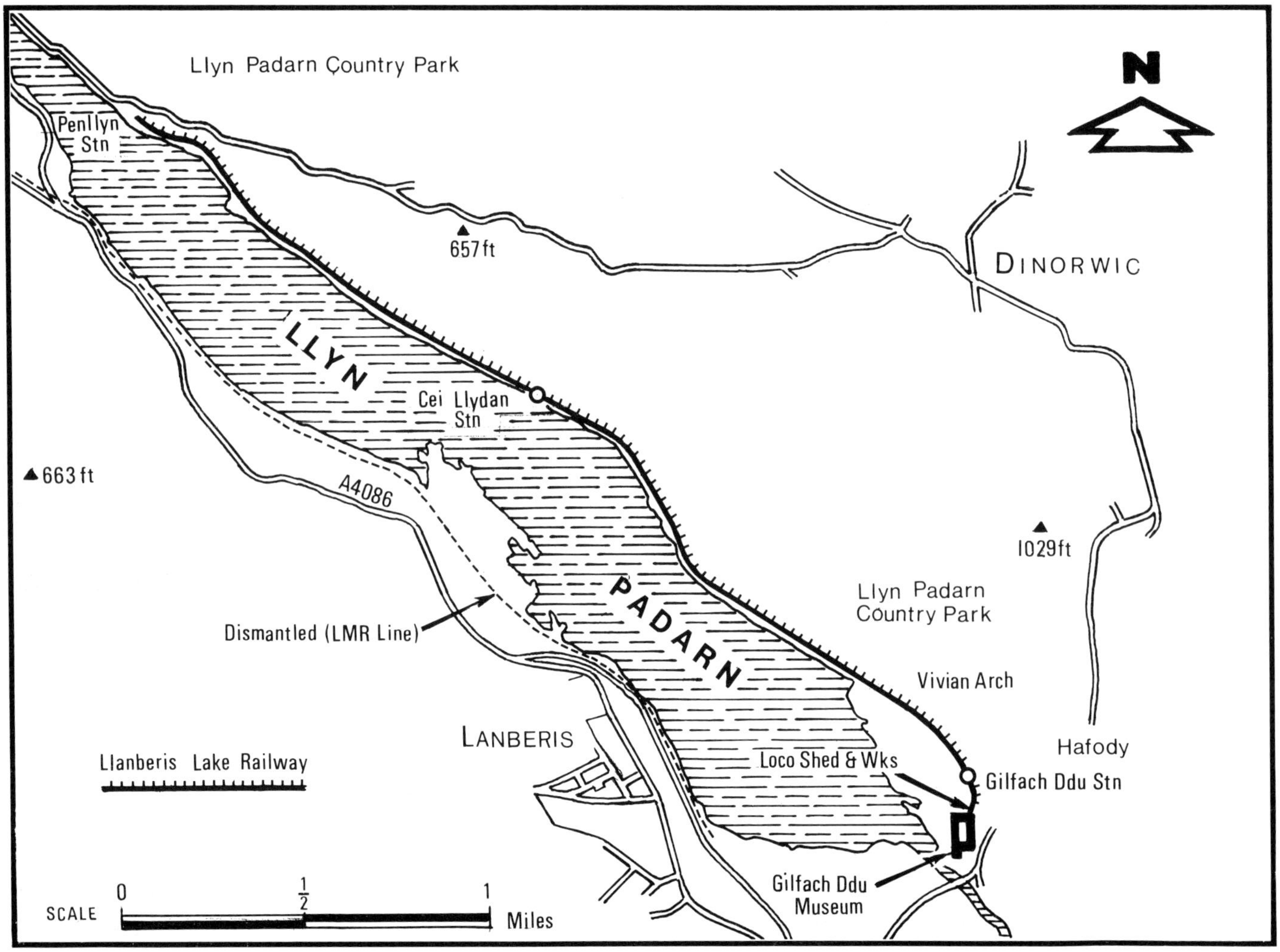

Bala Lake Railway

The construction of the Llanberis Lake Railway from 1970 onwards was then the only active scheme for a tourist railway alongside a Welsh lake. Some 45 miles to the south the attractions of the former GWR branch from Ruabon to Dolgellau were being considered for a new line, however, especially where it ran alongside Bala Lake.

The Bala & Dolgelly Railway was one of a chain of four independent railways, the others being the Vale of Llangollen, the Llangollen & Corwen and the Corwen & Bala Railways, which together ran westwards from Ruabon; the Bala & Dolgelly was opened in 1868. At Dolgelly there was an end on connection with the Cambrian Railways' branch from Barmouth Junction (now called Morfa Mawddach). The Great Western Railway took an interest in the promotion of these companies and in 1877 it absorbed the Bala and Dolgelly, the others following in 1896. During the summer the lines formed an important secondary route to the Welsh coast. The 1963 Beeching Report, *The Reshaping of British Railways*, recommended closure and this was implemented from 18 January 1965, the track being lifted in 1968.

The scheme to build a narrow gauge railway on the lakeside trackbed was made public in 1971. By this time the land was owned by the Merioneth County Council which proved to be amenable to the railway idea. A company, Rheilffordd Llyn Tegid Cyf, the first to be registered in Welsh, was formed to take a lease from the council and to operate the new line. Its base was established at Llanuwchllyn, a former passing place 4½ miles west of the old Bala Lake Halt. Stock and materials were sought and obtained; as at Llanberis, the gauge of the Bala Lake Railway, the Bala Lake Railway Co decided to encourage the involvement and support of volunteers; a supporters' society was formed and, after a false start, has proved to be a great asset to the company.

On 14 August 1972 the first train ran to a temporary terminus at Pentrepiod, 1½ miles from Llanuwchllyn; a notable feature of this section is the Ddolfawr bank, a half mile of 1 in 70 gradient in the Llanuwchllyn direction, a steep climb for the small locomotives which work the line. Initially services were worked by a small Ruston diesel hauling two open bogie coaches. A further three-quarters of a mile of track was laid and brought into use on 15 September the same year; 4,300 passenger journeys had been made by the time services finished for the winter.

The two open coaches used in the first season were built by Severn Lamb of Stratford-upon-Avon. MFD Engineering of Wrexham supplied three closed coaches to Llanuwchllyn, in 1973, 1974 and 1976, the last two having a guard's compartment. The remainder of the railway's fleet was locally built by Gwynedd Engineering Services in Bala; two closed coaches were delivered in 1978, followed by a similar, braked, vehicle in 1981 along with two semi-opens. The 1973-built carriage was withdrawn in 1979 and is being rebuilt as a mobile catering vehicle for stabling at Bala.

From 1973 until 1975 services were operated almost exclusively by a Severn Lamb Bo-Bo diesel hydraulic locomotive, named *Meirionnydd*. Equipped with a Leyland diesel engine, it outwardly resem-

Below:
At the western end of Llyn Tegid (Bala Lake) *Maid Marian* has nearly completed a round trip from Llanuwchllyn with a Bala Lake Railway train on 19 August 1990.

bles a member of British Rail's now-extinct 'Western' class of locomotives.

The year 1975 saw a further one mile extension, to Pantyr Hen Felin, brought into use at the same time as the railway introduced its first steam engine to the public. The locomotive was the 1903-built Dinorwic Quarry Hunslet 0-4-0T *Maid Marian* which had been in the ownership of the Maid Marian Locomotive Fund since 1965. Since then it had worked at Bessingham Gardens and on the Llanberis Lake Railway. Considering it has been hauling larger loads, longer distances, up steeper gradients than its designers could ever have envisaged it has been a great success in its third Welsh home.

Encouraged by *Maid Marian*'s success another Dinorwic Quarry Hunslet, 1902-built *Holy War*, was obtained, also in 1975; following an extensive overhaul at Llanuwchllyn it entered service in 1979. The frames of a third of these useful locomotives came to the railway in 1977. This one was 1902-built *Alice*, the second Dinorwic engine to have that name; reassembly commenced in 1987 but will not be completed until a new boiler is obtained. In 1980 yet another Dinorwic locomotive, *Jonathan* (formerly *Bernstein*) arrived at Llanuwchllyn; following repairs carried out in 1986 it went to the West Lancs Railway and in 1989 returned to the Lytham Creek Railway in Lancashire, having only worked running-in trips on the Lake railway.

In 1976 the railway was extended to Pen-y-bont, the site of Bala Lake Halt, the current terminus. Until 1988 little development took place here because it was intended to extend again, on a new formation which would take the railway around the end of the lake to a terminus on the Dolgellau road but in the winter of 1988/9 the platform was extended by three coach lengths, and a bay for a catering vehicle was provided. In pursuance of its extension plans the railway obtained The Rheilffordd Llyn Tegid Light Railway Order to cover the existing line on 9 October 1985. The National Park Bala area structure plan consultation documents published in November 1989 included a railway extension proposal for discussion by interested parties.

In 1977 a collection of privately owned diesel locomotives arrived from their increasingly vandalised home in Long Eaton, Derbyshire. They are on display with historical items of rolling stock, their owners forming themselves into the Llanuwchllyn Railway Museum Association. The working locomotives in the collection are sometimes used on works trains or volunteers' specials, a Saturday evening chip shop run to Bala being a regular excursion for them.

A further locomotive development took place in May 1979, when Kerr Stuart 'Wren' class engine, *Peter Pan*, was an attraction at the railway's gala weekend. Afterwards its owners took it back to England, leaving the Bala Lake's operators coveting an engine of its type. During 1980 negotiations took place with the owner of the 'Wren' in the Gloddfa Ganol collection at Blaenau Ffestiniog. Named *Ashover* by its owner, because of its somewhat tenuous links with the Ashover Light Railway, it was transferred to Llanuwchllyn in September 1980, where it became known as *Dryw Bach (Little Wren)* following a change in ownership; nameplates were not fitted in either case but were made for *Dryw Bach*. It was returned to steam in 1982 and proved to be a useful and popular engine. In 1985, however, it was found to be in need of major boiler repairs and was returned to Blaenau Ffestiniog.

In the winter of 1979 a number of alterations were made to the Bala & Dolgelly Railway station building at Llanuwchllyn. Secondhand bricks from Llanderfel and slates from Liverpool Exchange stations were used to ensure that the new works did not clash with the old. The canopy from Aberdovey was erected there at the same time, its second move for it had originally been part of the first station at Pwllheli.

The GWR signalbox at Llanuwchllyn was built in 1896, when the passing loop was installed, and remains in use. The original signalling installation was removed on closure so replacements have been found from several other sources. That so many of them came from the Lancashire & Yorkshire Railway is easily explained when it is known that the narrow gauge line's first general manager is a fan of that railway. The crossing loop at Llangower is also fully signalled and interlocked; a four-lever London & South Western Railway frame is used here.

Each year until 1978 the railway showed an increase in the number of passenger journeys made, 52,000 being the total for that year. By 1982 the figure had fallen to 40,000. To ease the company's financial situation the permanent staff was laid off for five months during the winter. A slight improvement in carryings took place in 1983 and it was able to retain its staff on a two-day week that winter; to further improve its situation by reducing borrowings and interest payments the company-owned cafe, close to the proposed Bala terminus, was sold. A further improvement in 1984 enabled the company to keep its staff until the end of 1986, when the establishment was reduced by two, to leave only the full-time general manager and part-time chief engineer on the strength. From that time the railway has depended mainly on its volunteers to meet all its operational requirements. For their part the volunteers set to with a will although it has been a long hard slog before an improvement in the railway's position was obtained; in 1988 passenger journeys, at 37,934, were 23% up on the previous year. A further traffic increase in the order of 10% was obtained in 1989.

The Bala Lake Railway clearly has a long way to go before it achieves the financial stability it requires but it does have the support of a hard-core of willing and hard working volunteers who will do everything they can to make it a success.

BIBLIOGRAPHY

Baughan, Peter E.; *A Regional History of the Railways of Great Britain Vol 11 North and Mid Wales*; David & Charles, 1980

Jackson, Chris & Briddon, Peter; *A visitor's guide to the Bala Lake Railway*; Cymmdeithas Rheilffordd Llyn Tegid, 1988

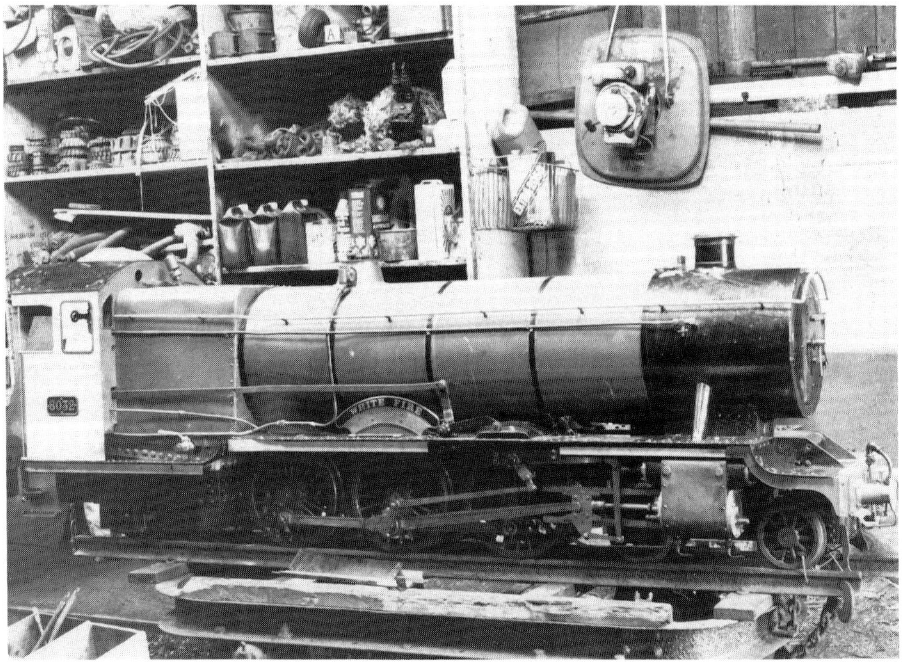

Right:
A contract job undertaken on the Bala Lake Railway was converting this Bullock 4-6-2 from 10¼in gauge to 7¼in gauge for owner Brian Hollingsworth. The locomotive was built in 1932 when it was called *The Monarch*. 19 August 1989.

WAGONS

Right:
A 'one-off' mixed-train ran on the Festiniog Railway on 1 May 1988 using vintage coaches and restored slate wagons and goods brake hauled by *Merddin Emrys;* the train is seen at Dduallt Tank Curve. *Norman Gurley*

Centre right:
On 19 August 1989 *Holy War* was steamed for test purposes and ran a works train to Bala, where the platform extension construction was under way.

Below left:
Restored Festiniog two- and three-ton slate wagons being unloaded at Pen-y-mount for their visit to the Welsh Highland Steam Gala in May 1988.

Below right:
The Snowdon Mountain Railway's works train is always the first working of the day, conveying stores and water to the Summit. It is seen returning to base on 18 August 1989.

Bottom:
During the Welshpool & Llanfair Railway's 1987 Steam Gala a goods train was operated and is seen arriving at Welshpool Raven Square Station behind *Joan.* *Hugh Ballantyne*

Right:
The oldest Festinog slate wagons had timber bodies; except for the one in the museum those remaining were scrapped in the 1950s. This one, photographed on 29 April 1989, is a recreation using parts disinterred from Glan-y-mor yard at Boston Lodge. Behind it is a four-wheeled brake No 2, built in 1885/6 by converting an 1865/6 quarrymen's coach of the same type as No 8, seen to the left. No 2's matchboard sides were restored for use on the mixed train in 1988.

Centre right:
The North Wales Quarrying Museum's *Una* works a goods train on the Llanberis Lake Railway on 19 August 1984. *Brian Dobbs*

Below left:
What is likely to be the last steam-hauled ballast train to be run under BR auspices ran on the Vale of Rheidol Railway in May 1988 and is seen here near Capel Bangor. *Andrew Bannister*

Below right:
The oldest working steam locomotive in regular use is the claim made for the Festiniog Railway's *Prince*, seen here shunting slate wagons at Boston Lodge on 3 November 1986.

Bottom left:
A mixed train run on the Welshpool & Llanfair Railway for the Industrial Railway Society on 23 September 1989. It was photographed at Heniarth, having just crossed the Banwy Bridge. *C. Shephard*

Bottom right:
The ultimate wagon train. Traditionally, loaded slate trains on the Festiniog Railway run by gravity. On 3 November 1986 the scene was recreated when a gravity train was run from Blaenau Festiniog to Porthmadog using restored laden slate wagons, and bogie Hudsons to provide additional braking power if required. To get it over the summit at Tanygrisiau Power Station the train was propelled thence by *Prince*.

Brecon Mountain Railway

The third new Welsh narrow gauge tourist railway was in the planning stages at the same time as the Llanberis Lake and the Bala Lake railways were being opened; like them it was built on an existing disused trackbed, alongside open water in a National Park. This line has the distinction of being in south instead of north Wales.

The Brecon & Merthyr Railway's route through the Brecon Beacons to Merthyr Tydfil was opened in 1963. Except for the industrial area of Merthyr the line ran through some of the most spectacular scenery in South Wales. This was to be of little use when the traffic declined so much that British Rail decided to close the line in 1964.

In 1972 the section of the Brecon & Merthyr route between Torpantau, Powis, and Pant, Mid Glamorgan, a distance of 5½ miles, was chosen for a new narrow gauge line to be called the Brecon Mountain railway; most of it was in the Brecon Beacons National Park where it ran alongside the Taf Fechan and Pentwyn reservoirs. By this time the trackbed had been sold to 14 different landowners. Following protracted negotiations settlements were made with 13 of them; the last owned the site of Pant station and the new company could not afford the asking price. The Brecon Mountain Railway (Light Railway) Order 1980 was enacted on 8 May of that year; in addition to authorising the construction and operation of the railway on the existing route, it also permitted the construction of a 600yd deviation on adjacent land at Pant. The new Pant station is actually sited on top of the disused Morlais Tunnel, with a smoke vent sited by the front door.

The first track was laid from a depot established at the former Pontsticill Junction to Pant, two miles to the south. 75lb rail, some of it from the Longmoor Military Railway in Hampshire, was laid to 60cm gauge. To speed up the operation, track panels were made up under cover at Pontsticill and carried to site by train.

Passenger services started on 8 June and four months later 17,000 passengers had been carried. The one-coach train was usually hauled by *Sybil*, an 0-4-0T Hunslet similar to the Dinorwic engines, but in this case originating from the Pen-yr-Orsedd Quarry in the Nantlle Valley, near Caernarfon. The carriage, soon to be followed by others, was built in the railway's own workshops at Pontsticill using a South African Railways bogie-wagon underframe as a foundation. With their end balconies and large windows these vehicles number amongst the smartest on the Welsh narrow gauge today, although the author well remembers the numbing effect of the seats!

The overseas involvement just mentioned with regard to the rolling stock is maintained with the line's motive power. In 1981 *Sybil* was joined in service by Arn Jung 0-6-2TWT *Graf Schwerin Lowitz* which came from the East German State Railways. Other locomotives awaiting restoration include a Baldwin Pacific and a Hanomag Garratt imported from South Africa and an Orenstein & Koppell 0-4-0WT, in need of a new boiler, from Germany.

Other steam locomotives are owned by the company for restoration as museum exhibits; they include two with vertical boilers, one from Pen-yr-Orsedd and the other once owned by the Penmaenmawr granite quarry owner for use in his garden railway. Three diesel locomotives and a Wickham trolley are available for engineering department use or in an emergency. One of the diesels is an 0-6-0 diesel hydraulic locomotive built by the railway 1987 using components made by Baguley-Drewry Ltd of Burton-on-Trent for an export order which was cancelled. The Brecon Mountain Railway acquired the parts after Baguley-Drewry went into liquidation and in addition to building their own locomotive built a second for British Rail to use on the Vale of Rheidol Railway.

Facilities at Pontsticil are of a temporary nature, partly because of its position part way along the proposed completed railway and partly because the National

Below:
A Brecon Mountain Railway train heads for Pant at midday on 6 August 1989. The railway's entire carriage stock is seen.

Right:
The Brecon Mountain Railway built a second locomotive to an order from British Rail and delivered it to Aberystwyth for use on the Vale of Rheidol Railway where it became No 10 on 21 June 1988. *Mike Haddon*

Park authority, in trying to limit the amount of traffic in the park, insists that all journeys commence outside its boundaries. A good job has been made of converting some ex-British Rail covered carriage trucks into a cafeteria, waiting area and conveniences. For those who do not wish to return immediately to Pant there is a network of footpaths which start from the station by which they can explore the locality.

Permanent facilities have been provided at Pant, where what is arguably the most impressive narrow gauge station building to be built in many years is now in use. Given an appearance reminiscent of a Welsh chapel externally, it incorporates a well equipped machine shop and erecting shop for the maintenance and repair of locomotives and rolling stock and a restaurant in addition to the usual passenger facilities; the layout is such that passengers pass a workshop viewing gallery whilst en route between booking hall and platform. One of the first tasks undertaken here was the construction of a caboose modelled on a Sandy River & Rangeley Lake Railroad original which entered service in 1990; it is heated, by a stove, and is capable of carrying wheelchair-bound passengers. The local authority provided the large station car park.

Materials are being collected for the Torpantau extension but both locomotive and carriage fleets will have to be extended before the railway can cope with the additional traffic likely to be generated by it. When completed, passengers will be able to enjoy riding through forest and the 666yd tunnel at Torpantau before emerging at the northern terminus Glyn Collwyn. At 1,313ft above sea level the tunnel will be the highest in Britain. An interesting facet of operating this section will apply where the railway passes through the Tal Fechan Forest. It arises from a requirement of the Light Railway Order, that use of any motive power that is not oil-fired steam, internal combustion or electric must be approved by the Forestry Commissioners.

As already described, in 1989 the Brecon Mountain Railway Ltd became the owner of the Vale of Rheidol Railway; Vale of Rheidol locomotives are likely to be fairly regular visitors to Pant for overhaul.

BIBLIOGRAPHY

Johnson, Peter; 'The Brecon Mountain Railway – South Wales' Great Little Train'; *Railway World*, November 1985.

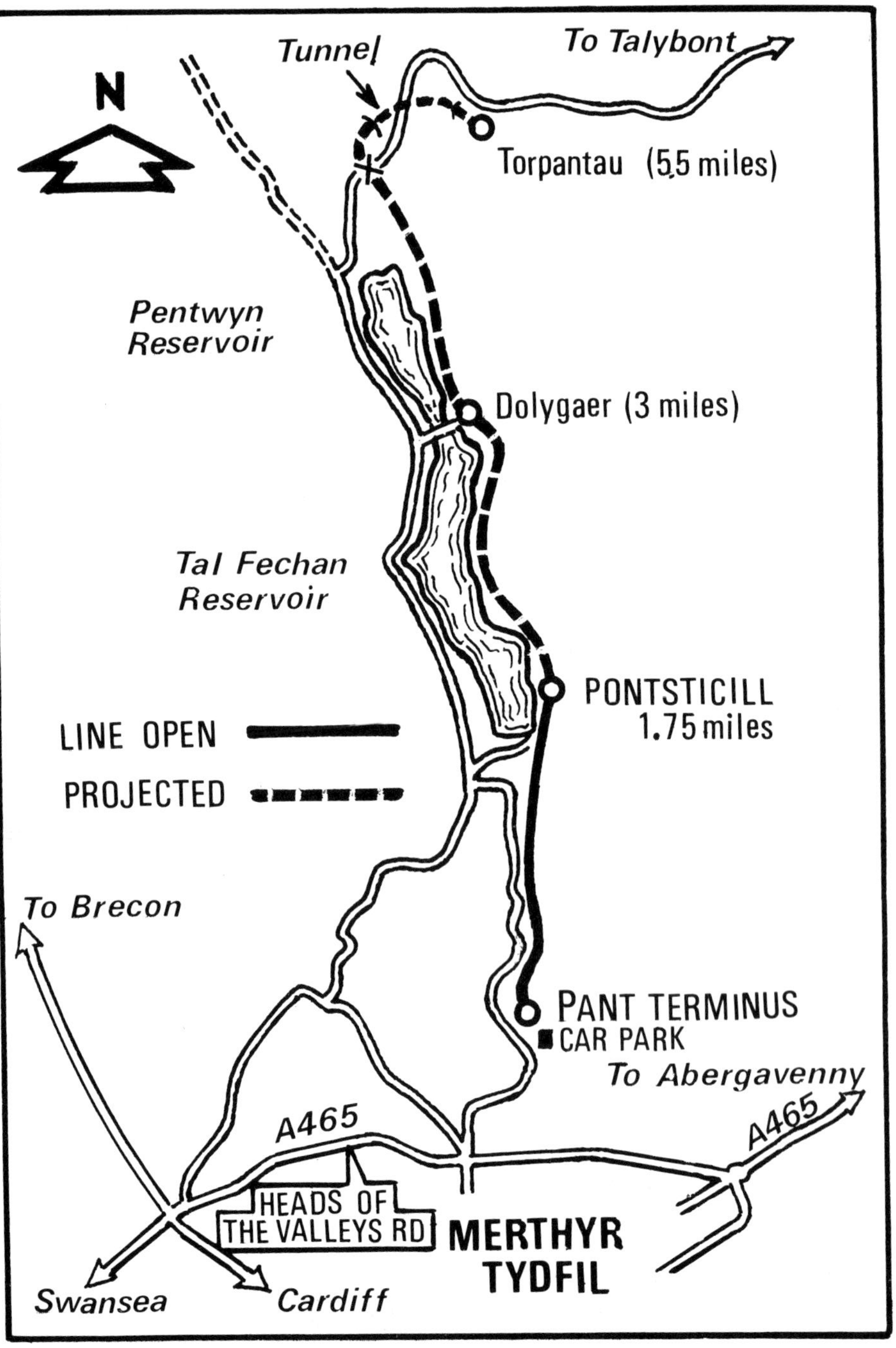

STATIONS

Right:
Devil's Bridge, the inland terminus of the Vale of Rheidol Railway is 680ft above sea level. It is seen here on 18 June 1983.

Centre right:
The old, 15in gauge, station at Fairbourne seen outlined by the arch of the carriage shed. The locomotive, *Ernest W. Twining* was shipped to Japan in 1987. September 1983.

Below left:
The Llanberis Lake Railway's *Dolbadarn,* left, and *Elidir* cross at Cei Llydan in August 1986.

Bottom left:
The old Fairbourne Railway used to serve a golf course, now long disused with a platform called Golf Halt. In 1984 the new owners renamed it Gorsafawwdacha' idraigddanheddollon-penrhynareurdaethceredigion as being more discriptive of the location! It means 'the Mawddach station with its dragon's teeth on North Penrhyn Drive by the golden sands of Cardigan Bay' and contrives to produce the longest place name in the world. 15 August 1989.

Bottom right:
At Minffordd the Festiniog Railway's station predates British Rail's Cambrian Coast station of that name which is served by a subway. On 22 May 1987 *Hinton Manor* passes underneath *Blanche* **and the narrow gauge line when running a gauging trip to Pwllheli prior to operating a number of excursions there that year.**
Norman Gurley

Top left:
Two trains at Clogwyn Station on 7 April 1977. During the early part of each year Snowdon Mountain Railway trains terminate here due to snow filling the cuttings as shown. *Ray Ruffell*

Top right:
Talyllyn Railway No 4 *Edward Thomas* waits for departure time on a quiet Sunday morning at Tywyn Wharf Station. Notice the photographer with a plate camera on the left. May 1984.

Above:
Porthmadog Welsh Highland Railway station, alongside the BR Cambrian Coast line on 8 September 1984. *Tom Heavyside*

Above right:
Tan-y-bwich is the largest and busiest of the Festiniog Railway's intermediate stations and was a terminus between 1958 and 1968. *Merddin Emrys* was running in the black livery it carried between its recent overhaul and being painted in 1880s livery on 30 December 1987.

Right:
Until 22 May 1976 Abergynolwyn was the Talyllyn Railway's passenger terminus. The opening of the Nant Gwernol extension on that date saw the timetable modified so that Down trains stop long enough to allow passengers to use the facilities there on 21 May 1983.

Welsh Highland Railway

The Welsh Highland Railway which now solicits for passengers at the opposite end of Porthmadog to the Festiniog Railway has its roots in four other 1ft 11½ lines but has no physical connections with any of them. Its supporters do have great hopes for the future though.

The earlier systems were: the Croesor Tramway, a horse tramway which gave the Croesor Valley slate quarries access to Portmadoc; the North Wales Narrow Gauge Railways, a railway line with a main line between Dinas Junction, near Carnarvon, and South Snowdon and a branch to Bryngwyn; the Portmadoc, Beddgelert & South Snowdon Railway, a company which had powers to connect the NWNGR to the Croesor Tramway, via Beddgelert and the Aberglasyn Pass, to upgrade part of the Croesor, and to electrify the entire 21 mile route – the remains of incomplete and abandoned PB&SSR construction around Beddgelert may still be seen; and the Welsh Highland Railway (Light Railway) of 1922. The latter was created to take over the NWNGR, which had been moribund since 1916, and the PB&SSR, which had ceased trading in 1919, and to restore and complete them as appropriate. The electrification plans were abandoned, resulting in a new route suitable for steam locomotives being constructed around Beddgelert.

The Welsh Highland Railway (Light Railway) scheme was authorised by Light Railway Orders obtained in 1922 and 1923; a Festiniog Railway Light Railway Order of 1923 sanctioned, *inter alia,* a link with the WHR at Portmadoc. The complete railway was opened in 1923.

The new railway owned three steam locomotives, *Moel Tryfan,* a NWNGR 0-6-4T single Fairlie built by Vulcan Foundry in 1875 but since 1917 incorporating the frames of the identical *Snowdon Ranger; Russell,* a Hunslet 2-6-2T built for the PB&SSR in 1906; and an ex-War Department Baldwin 4-6-0PT built in 1917. The WHR's coaching stock was the surviving NWNGR fleet of 11 bogie vehicles.

The FR link enabled through services to run from Blaenau Ffestiniog to Dinas but the interworking of stock was not easy; the brakes and couplings were incompatible and the FR's loading gauge too small for WHR stock. *Russell* was taken into Boston Lodge for cutting down to fit; the result was not a success, either visually or physically.

Losses made each year resulted in a receiver being appointed in 1927. When no financial improvement was found to be possible during the following six years, the line was closed. In the following year, 1934, it was reopened by the FR under the terms of a 42-year lease. Despite attempts to increase traffic the FR could do no better and it was decided not to operate the line from 1937. The Welsh Highland Railway was abandoned to the elements.

In 1941 the track between Dinas and Croesor Junction was, with other material and equipment, requisitioned to aid the war effort. The remainder was left in case the Croesor quarries reopened when hostilities ceased. They did not so. Most

Below:
The Welsh Highland Railway's Gelerts Farm Works appears in view for the passengers of this *Karen*-hauled train returning to Porthmadog.

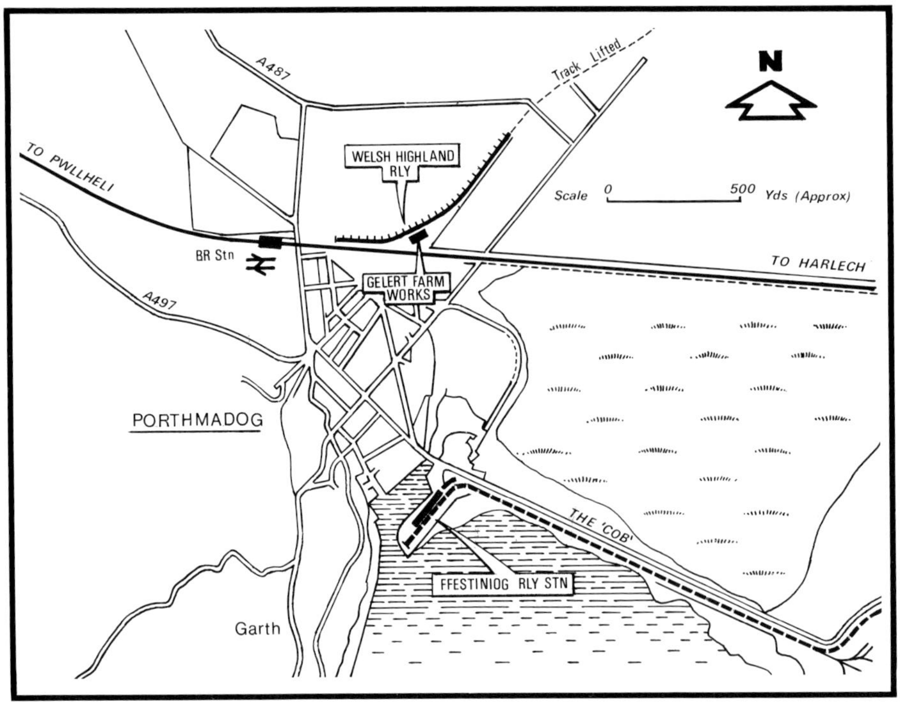

of the remaining track, including that of the Croesor Tramway was lifted in 1950.

The scheme to revive the Welsh Highland Railway began in 1961, with the formation of a society. Sufficient progress was made that in 1964 it was incorporated as the Welsh Highland Light Railway (1964) Ltd, a company limited by guarantee. Unfortunately the 1922 company's liquidator, with whom the 1964 company had agreed trackbed purchase terms, died before contracts could be exchanged. The Official Receiver took over the liquidator's affairs and very little progress had been made over the intervening years, although the 1964 company has reached an agreement with Gwynedd County Council that has the Council negotiating with the Receiver to buy the trackbed which it will then lease to the 1964 company.

While negotiations continued the company set about acquiring locomotives and rolling stock, storing them first at a base in Kinnerley, on the former Shropshire & Montgomery Railway. Its members were especially pleased when they received the only surviving WHR locomotive, *Russell,* as a donation in 1965.

In 1973 the company gained access to the Beddgelert exchange siding, at Porthmadog, and alongside the WHR trackbed and the British Rail Cambrian Coast line, on which it has built a ¾-mile long railway. In 1975 it brought the adjoining Gelert's Farm at which a depot has been established. Authority for the railway was obtained in a Light Railway Order effective from 12 May 1980.

Public services commenced on 2 August 1980, using a four-wheel diesel and a homemade coach. Steam was used for the first time on 30 April 1983 when *Karen,* a 1942-built Peckett 0-4-2T which had spent its working life in Rhodesia, entered service after a seven-year overhaul.

In addition to establishing an operational railway, however, the company's greatest achievement must be the restoration of *Russell* to working order. As early as 1970 a new boiler had been fitted by the makers but it was not until 1984 that restoration commenced in earnest, a task which took a further three years; fitted with many new components, *Russell* worked its first passenger trains on 4 April 1987. The following year it returned triumphantly to the Festiniog Railway, where it ran passenger trains between Porthmadog and Rhiw Goch as part of the FR's 125th Anniversary of Steam celebrations.

BIBLIOGRAPHY

Boyd, J. I. C.: *Narrow Gauge Railways in South Caernarvonshire;* Oakwood Press, 2 vols 1988/9.

Above right:
With Moel-y-gest forming a backdrop, the Welsh Highland Railway's *Russell* runs round its train at Pen-y-mount. The locomotive carries the name of the Portmadoc, Beddgelert & South Snowdon Railway's Chairman and Receiver.

Right:
A mixed scene in the Welsh Highland Railway's Gelerts Farm locomotive shed in May 1988. Included are WHR Transport Gala visitors *Chaloner* and *Britomart*, the event being held at the end of that month, the diesel is *Cnicht*, a Motor Rail built in 1941 and painted in the colours of the Yorkshire Water Authority, its former owners.

FESTINIOG RAILWAY 150TH ANNIVERSARY

Above:
Prince **at Blaenau Ffestiniog with a special train for the Conwy Valley Sprinter launch on 24 March 1986, the first event of the 150th anniversary year.**

Right:
The Festiniog Railway was opened on 20 April 1836. In 1986 20 April fell on a Sunday so a gala weekend was arranged. *Britomart* and *Chaloner* spent the weekend hauling slate wagons up Gwyndy Bank and are seen here in the Minffordd mineral line preparing to set off again on 19 April.

Bottom left:
Having got the wagons up to Gwyndy Bank they were released to return by gravity, the first occasion a gravity train had been publicly demonstrated since the Centenary of Steam in 1963, and then only four wagons had been used!

Bottom right:
The last event of the year occurred on 3 November when another, longer, gravity train was run, starting at Blaenau Ffestiniog! The train is seen arriving at Blaenau with *Prince* in charge.

Teifi Valley Railway

The Teifi Valley Railway is situated in South West Wales, north of Carmarthen, occupying part of the trackbed of the nine-mile long Newcastle Emlyn branch in the county of Dyfed.

The route has its origins in the broad gauge Carmarthen & Cardigan Railway which opened its line from Carmarthen as far as Llandyssul in 1864. In 1866 the Manchester & Milford Railway, building a standard gauge line to Aberystwyth, connected with the C&CR at Pencader, three miles from Llandyssul; a third rail was laid between the junction and Carmarthen by the end of that year. The Carmarthen & Cardigan Railway was converted to standard gauge in 1872 and the company was leased by the Great Western Railway in 1881. Nothing ever came of the Cardigan scheme but the GWR did take advantage of C&CR powers to extend the line to Newcastle Emlyn in 1895. From 1906 the GWR absorbed the Manchester & Milford Railway and the Newcastle Emlyn line became a branch of the Aberystwyth-Carmarthen main line.

The passenger service was withdrawn in 1952, followed by the pick-up freight in 1965 and the through freight in 1973; the last loads were coal and fertiliser. A Teifi Valley Railway Preservation Society had already been formed, in 1972, intending to buy the Carmarthen-Newcastle Emlyn line after closure and to run it as a standard gauge railway. This scheme was brought down by the rampant inflation of the time and the society eventually turned its attention to what is now the Gwili Railway. Some members were still interested in preserving part of the Newcastle Emlyn line and negotiations concerning the branch soon commenced. In 1978 the Vale of Teifi Narrow Gauge Railway Society was formed, for narrow gauge was to be the future for this railway. A Light Railway Order was obtained by the Dyfed Railway Co, the operating company, on 15 December 1980 and in 1981 the formation was purchased from British Rail. The Order sanctioned the construction of a railway line with a gauge of 1ft 11¾in or 60.325cm.

A base was established at Henllan station and the first visible progress was made when a Manpower Services Commission team started work in 1983. The MSC cleared the undergrowth, built a toilet block and laid 800yd of track westwards from Henllan to Prontprenshitw; the Henllan platform is to the west of the

Below:
1894-built Hunslet *Alan George* backs onto its train at Henllan, the Teifi Valley Railway's centre of operations, on 13 August 1989.

Right:
The driver of the Teifi Valley Railway's *Alan George* prepares to set back onto his train at Henllan on 13 August 1989. Unlike many former quarry locomotives working in preservation, this one has not been fitted with a cab – Sir Arthur Heywood's dictum of providing the driver with a 'stout mackintosh' also applies! Owned by a group of supporters, the locomotive's boiler has been retubed and fitted with a new firebox at Henllan. The track beyond the gates runs through the former standard gauge station to the Teifi's stock shed.

standard gauge station because of a Department of Transport requirement that it should be on the level. Two further MSC schemes saw to the creation of a car park and a children's playground in the old goods yard and a mile-long nature trail and an open air theatre in the woods alongside the track.

The first passenger train, consisting of Hunslet diesel locomotive *Sholto* and one coach, left Henllan at 1pm on 24 August 1985. Operating at weekends until the end of September, 905 passengers were carried.

The official opening took place on 9 April 1986, Peter Prior performing the honours. In its first full year of operating the little railway carried 6,521 passengers in its diesel-hauled train. On 12 April 1987 steam was introduced in the form of *Alan George,* an 1894 Hunslet 0-4-0T from the Penrhyn Slate Quarry owned by a group of supporters. The railway has four carriages, three of them built, at Henllan, on Baguley-built underframes of former Trentham Gardens Railway toast-rack carriages; the fourth was built, also at Henllan, on an underframe supplied to order by the Talyllyn Railway.

The extension to Llandyfriog, opened in 1989 and giving a total run of 1,800yd was financed by a bond issue; by this time the operating company had obtained PLC status, becoming the Teifi Valley Railway PLC at the same time. In the long term the railway has ambitions to extend both westwards to Newcastle Emlyn, where it intends to deviate from the original station by bridging the Teifi and terminating by the castle, and eastwards to Llandyssul, where it will be close to the Carmathen-New Quay main road. At its ultimate the line will be six miles long; the trackbed between Llandyssul and Pencader was disposed of as surplus to any foreseeable requirements in 1988.

The railway intends to provide more than just a train ride and passengers are positively encouraged to leave the train at Pontprenshitw to examine the fine underbridge which carries the railway over the ravine of the Afon Cynllo; the amphitheatre and nature trail are served by Forest Halt, 200 metres on the Henllan side of Pontprenshitw, the amphitheatre being the location for the entertainments during the high season. The railway's staff are encouraged to be extrovert and are instructed to talk to the passengers, to ensure that passengers depart feeling thoroughly entertained.

BIBLIOGRAPHY

Padfield, Roger & Burgess, Barrie; *The Teifi Valley Railway;* Laidlaw-Burgess (Publishers), 1974

TVR; *Teifi Valley Railway Tourist Guide;* Teifi Valley Railway plc, 1989

Right:
The Teifi Valley Railway's *Sholto* is a four-wheeled Hudson Hunslet built in 1941. It worked all trains in 1986 and many in 1987. The locomotive on the train is a 1959-built Simplex, *Sammy*, which has only been on the railway for six months when photographed on 13 August 1989. On this occasion it was providing continuous braking to the train due to the failure of the steam locomotive's air pump.

FESTINIOG RAILWAY 125TH ANNIVERSARY OF STEAM

Right:
This anniversary was also commemorated by a gala weekend, this time taking advantage of a public holiday to make a three day event. *Welsh Pony* built in 1867, was withdrawn in 1938 and has been displayed outside Harbour Station for several years; it is now bearing its 1900s livery.

Centre left:
Several local ladies turned out in Welsh national dress and are seen with Alan Pegler, the Company's President.

Centre right:
The Welsh Highland Railway's *Russell* was a popular visitor and is seen here at Harbour Station.

Bottom left:
***Merddin Emrys* and the mixed train leaving Tan-y-Bwlch on 1 May.**

Bottom right:
***Prince* with commemorative headboard and vintage stock arriving at Minffordd.**

Fairbourne and Barmouth Steam Railway

The two-mile long Fairbourne Railway started life as a 2ft gauge horse tramway in the 1890s. It was built to aid the development of Fairbourne as a seaside resort, intended to compete with the nearby Barmouth. Much of the development was undertaken by McDougall of 'self raising' flour fame. The tramway followed the old road to Penrhyn Point, the embarkation point for ferries across the Mawddach estuary to Barmouth.

As the housing became available for occupation passenger cars were introduced on the tramway; it ceased to carry materials when building finished. In 1916 the line was taken over by Narrow Gauge Railways Ltd (then owners of the Ravenglass & Eskdale Railway) who, in conjunction with Bassett-Lowke, the Northampton model builder, converted it to 15in gauge and introduced steam traction. In this form the railway continued through two changes of ownership until the last war.

After the war the line was in a severely damaged state and unfit for use. Ownership changed again and it was rebuilt, opening throughout in 1947. In the 1950s and 1960s increasing passenger traffic encouraged the owners to make a considerable investment in new locomotives and rolling stock.

As much of the line runs along the dunes it is very much at the mercy of the sea and shifting sand. By the 1970s the seaward terminal had ceased to be readily accessible from the beach used by the ferries so a deviation line was built to overcome this problem, taking the line to the landward side of the Point and coming into use in 1976. Its first passenger use was actually on 22 May, a non-service day but the same day that the Talyllyn Railway opened its Nant Gwernol exten-

Below:
A typical 12¼in gauge view of the Fairbourne & Barmouth Railway, with the half size reproduction Lynton & Barnstaple Railway 2-6-2T *Yeo* and train of Frestiniog outline coaches at Penrhyn Point on 15 August 1989.

Below:
After the Fairbourne Railway changed hands in 1984 a number of changes were soon implemented, even whilst the line continued operating at 15in gauge; notably, *Sian* was converted to American outline and renamed *Sydney*. *Peter Groom*

Above:
The half size Himalaya-Darjeeling 'B' class 0-4-0 tank, *Sherpa*, enters the Fairbourne & Barmouth Railway's passing loop, where a crossing will take place with Lynton & Barnstaple type 2-6-2T *Yeo.*

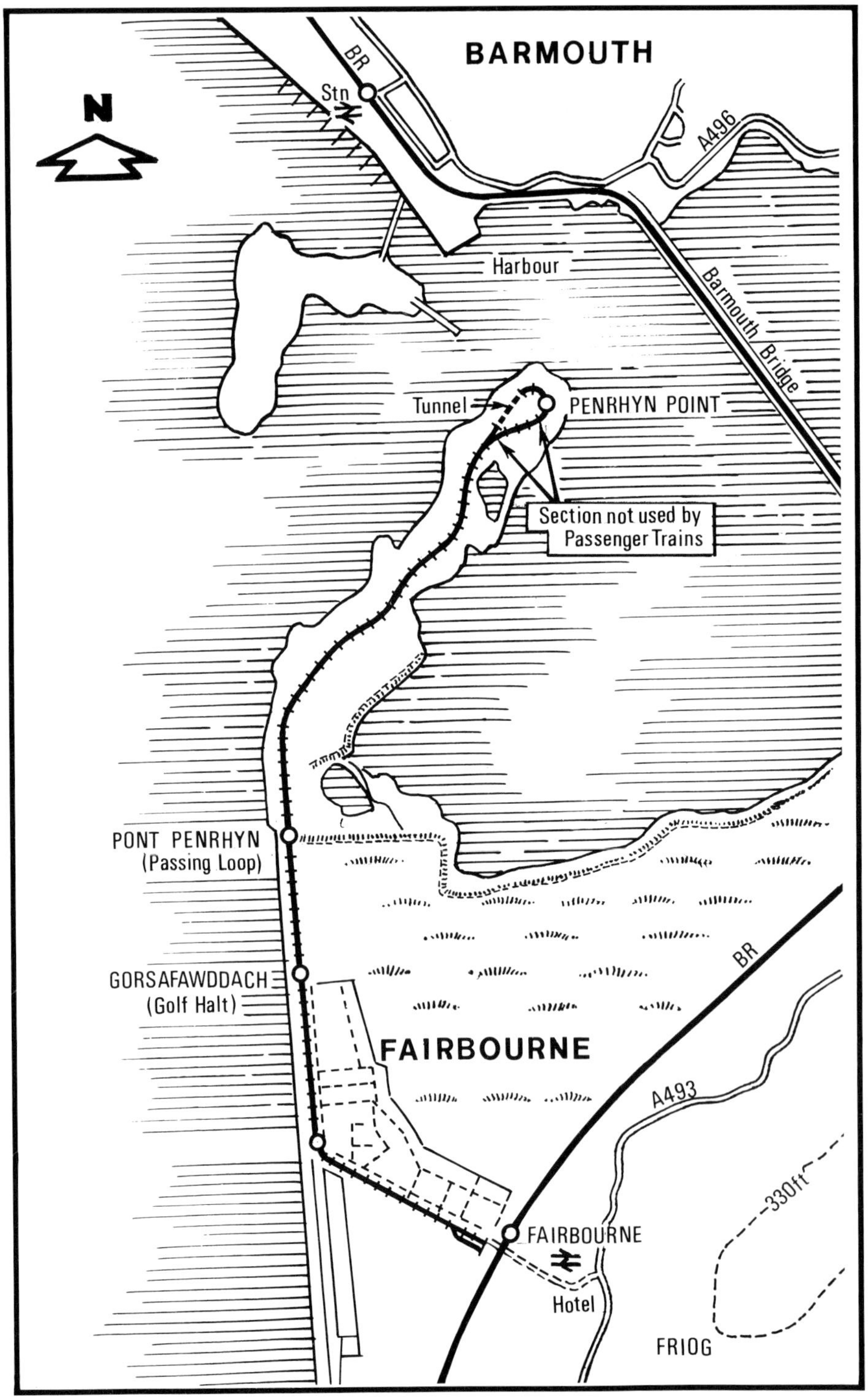

sion; so a railtour party from Leicester managed to ride on two new narrow gauge extensions on one day!

In 1984 the railway changed hands again, bringing new owners who made changes to the line. Chief amongst these are the regauging of the track to 12¼in gauge, the realigning of the track (including construction of a tunnel) at the Point and the provision of new terminal buildings at both Fairbourne and the Point. At the Point the new route again takes the line seawards, joining up with the old route at the station and giving the appearance of a terminal loop, as at Dungeness on the Romney, Hythe & Dymchurch Railway, but here the engines run-round at the station and trains leave in the same direction as they arrived.

To work the regauged railway four steam engines built in England in 1978/9 for a French line were provided; they were half sized replicas of prototypes from the Lynton & Barnstaple, North Wales Narrow Gauge, Leek & Manifold and the Himalaya-Darjeeling narrow gauge railways.

Whilst preparations were put in hand for the gauge conversion 15in gauge operations continued in 1985, although with some changes. Converted to American outline, 2-4-2 *Sian* ran as *Sydney* and the petrol hydraulic *Sylvia* became the diesel hydraulic *Liilisn Walter,* again taking an American outline, and later being converted to 12¼in gauge for the 1986 season. Some of the 12¼in gauge carriages, also from the French line, ran on 15in bogies during 1985. Later, *Sydney* and other 15in gauge stocks were sold.

Operation of the 12¼in gauge trains began on 28 March 1986 but it was not the end of development at Fairbourne. In the workshops the Leek & Manifold outline 2-6-4T was extensively rebuilt to become *Russell,* although still a 2-6-4T instead of the 2-6-2T arrangement of the prototype, a new battery powered *Gwiril* was built in 1987 and work started on the construction of an American 2-6-2 inspired by a Sandy River & Rangeley Lakes Railway prototype; a butterfly farm was established at Fairbourne and at the Point Breton pancakes are the main attraction at the cafe there. From 1988 the line has been known at the Fairbourne & Barmouth Railway. In 1990, after celebrating its centenary, the railway was put on the market, with an asking price of £1.5 million.

BIBLIOGRAPHY

Smithers, M.; 'Slim Gauge from Fairbourne'; *Railway Magazine,* October 1988

Other Narrow Gauge Railways and Museums

Great Orme Tramway
A cable car system to the Great Orme Summit at Llandudno.

Rhyl Marine Lake Miniature Railway
15in gauge circular system first opened in 1911. Now opens on fine days during the summer.

Narrow Gauge Railway Museum
Situated at the Talyllyn Railway's Wharf Station, Tywyn.

Festiniog Railway Museum
Sited at the Festiniog Railway's Harbour Station, Porthmadog.

North Wales Quarrying Museum
Located at Gilfach Ddu, Llanberis. Locomotive operates on some days in high summer.

Industrial Locomotive Museum
Situated at Penrhyn Castle, Bangor. A National Trust property.

Conwy Valley Railway Museum
Sited at Betws-y-Coed Station. Operates a 7¼in gauge miniature railway and a 15in gauge tramway.

Llechwedd Slate Caverns, Blaenau Ffestiniog
Operates a 2ft gauge railway into the caverns.

Gloddfa Ganol, Blaenau Ffestiniog
Operates a 2ft gauge railway around the site. Also accommodates a collection of narrow gauge exhibits.

Corris Railway Museum
Located at Corris. In addition to the Museum the Society intends to recreate part of the Corris Railway as far as Maespoeth.

Top right:
The 3ft 6in gauge Great Orme Tramway operates in the street, San Francisco style, on the lower section, Car 5 is seen on Tygwyn Road on 16 August 1989.

Centre right:
The Conwy Valley Railway Museum operates an extensive 7¼in gauge line at Betws-y-Coed. The Denver & Rio Grande 2-8-0 *Old Rube* was the centre of attraction when viewed from the BR station on 16 August 1989.

Right:
The Narrow Gauge Railway Museum, to be found at the Talyllyn Railway's Wharf station, includes eight steam locomotives in its display; ex-Dinorwic Quarry Hunslet was photographed on 14 August 1989.

Above:
The battery locomotive-operated Llechwedd Slate Cavern tour contines to be extremely popular with visitors to Blaenau Ffestiniog. In April 1984 one of the specially constructed trains emerges into the open on a part of the route which has the roof removed to give access to the good slate for which there is still a ready market.

Top right:
With very little publicity a narrow gauge railway was opened on the Gloddfa Ganol site of the former Oakeley Quarry, Blaenau Ffestiniog, on 28 July 1987, running between the shop and mine entrance. Motor Rail 40s 308 is seen working the line on 17 August 1989; the locomotive is one of three available for traffic which formerly belonged to the Severn Trent Water Authority.

Centre left:
The Corris Railway Society owns both the former Corris Railway stable block at Corris and the locomotive shed at Maespoeth, seen here looking towards Corris on 29 April 1984. Three-quarters of a mile of 2ft 3in gauge track has been laid between the two points, and the Society does hope to operate a passenger service.

Centre right:
The North Wales Quarrying Museum at Gilfach Ddu, Llanberis, was formerly the Dinorwic Quarry Workshops. Their ex-Pen-yr-orsedd Slate Quarry Hunslet 0-4-0T *Una* is kept in working order and is sometimes steamed for the benefit of visitors. Photographed in August 1986, the backdrop is the former Vivian Quarry.

Right:
Operation of the Rhyl Miniature Railway is usually restricted to the peak summer season and by the weather. On a fine 7 July 1987 *Joan* takes water before working the last circuit of the day. The locomotive was built in Rhyl for this railway to the design of noted miniature-railway engineer, Henry Greenly. *Larry Goddard*